STUDIES IN APPLIED MECHANICS 25

Convex Models of Uncertainty in Applied Mechanics

STUDIES IN APPLIED MECHANICS

1. Mechanics and Strength of Materials (Skalmierski)
2. Nonlinear Differential Equations (Fučík and Kufner)
3. Mathematical Theory of Elastic and Elastico-Plastic Bodies. An Introduction (Nečas and Hlaváček)
4. Variational, Incremental and Energy Methods in Solid Mechanics and Shell Theory (Mason)
5. Mechanics of Structured Media, Parts A and B (Selvadurai, Editor)
6. Mechanics of Material Behavior (Dvorak and Shield, Editors)
7. Mechanics of Granular Materials: New Models and Constitutive Relations (Jenkins and Satake, Editors)
8. Probabilistic Approach to Mechanisms (Sandler)
9. Methods of Functional Analysis for Application in Solid Mechanics (Mason)
10. Boundary Integral Equation Methods in Eigenvalue Problems of Elastodynamics and Thin Plates (Kitahara)
11. Mechanics of Material Interfaces (Selvadurai and Voyiadjis, Editors)
12. Local Effects in the Analysis of Structures (Ladevèze, Editor)
13. Ordinary Differential Equations (Kurzweil)
14. Random Vibration—Status and Recent Developments (Elishakoff and Lyon, Editors)
15. Computational Methods for Predicting Material Processing Defects (Predeleanu, Editor)
16. Developments in Engineering Mechanics (Selvadurai, Editor)
17. The Mechanics of Vibrations of Cylindrical Shells (Markuš)
18. Theory of Plasticity and Limit Design of Plates (Markuš)
19. Buckling of Structures—Theory and Experiment. The Josef Singer Anniversary Volume (Elishakoff, Babcock, Arbocz and Libai, Editors)
20. Micromechanics of Granular Materials (Satake and Jenkins, Editors)
21. Plasticity. Theory and Engineering Applications (Kaliszky)
22. Stability in the Dynamics of Metal Cutting (Chiriacescu)
23. Stress Analysis by Boundary Element Methods (Balaš, Sládek and Sládek)
24. Advances in the Theory of Plates and Shells (Voyiadjis, Karamanlidis, Editors)
25. Convex Models of Uncertainty in Applied Mechanics (Ben-Haim, Elishakoff)

STUDIES IN APPLIED MECHANICS 25

Convex Models of Uncertainty in Applied Mechanics

Yakov Ben-Haim

Faculty of Mechanical Engineering, Technion - Israel Institute of Technology, Haifa, Israel

and

Isaac Elishakoff

Center for Applied Stochastics Research and Department of Mechanical Engineering, Florida Atlantic University, Boca Raton, Florida, U.S.A.

ELSEVIER

Amsterdam — Oxford — New York — Tokyo 1990

ELSEVIER SCIENCE PUBLISHERS B.V.
Sara Burgerhartstraat 25
P.O. Box 211, 1000 AE Amsterdam, The Netherlands

Distributors for the United States and Canada:
ELSEVIER SCIENCE PUBLISHING COMPANY INC.
655 Avenue of the Americas
New York, N.Y. 10010, U.S.A.

Library of Congress Cataloging-in-Publication Data

Ben-Haim, Yakov, 1952-
Convex models of uncertainty in applied mechanics / Yakov Ben-Haim and Isaac Elishakoff.
p. cm. -- (Studies in applied mechanics ; 25)
Includes bibliographical references.
ISBN 0-444-88406-8
1. Convex sets. 2. Probabilities. 3. Mechanics, Applied.
I. Elishakoff, Isaac. II. Title. III. Series.
QA640.B45 1990
620.1--dc20 89-77748
CIP

ISBN 0-444-88406-8 (Vol. 25)
ISBN 0-444-41758-3 (Series)

Printed in The Netherlands.

To
Miriam, Zvika, Eitan and Rafi
Y.B.-H.

To
Esther, Ben and Orly
I.E.

Foreword

Professors Ben-Haim and Elishakoff honored me with their request that I add some introductory remarks to those they make themselves in this book, and make better. It is perhaps disquieting that any such remarks should be desirable. Their book is timely, even overdue, and should be widely welcomed. It treats a problem that recurs in most applications of mathematics to the real world, namely that of how to accommodate uncertainties in the empirical and theoretical information which underlies such applications.

That an accommodation is absolutely necessary has been recognized for a long time in the natural sciences as well as in engineering, and the traditional way of achieving it has been to equate uncertainty with probability. It is an idea that was, without question, spectacularly successful. In physics, it produced statistical mechanics and quantum theory. It entered into engineering perhaps mainly through the work of S.O. Rice on noise in electronic devices and of N. Wiener on radar fire control, and it led to C.E. Shannon's information theory, among others.

Not all of these successes were unqualified, however, and engineers in particular did not accept them without questioning. Chap. 1 of this book quotes from a large literature that documents the soul-searching that went on. (The development by L.A. Zadeh of the theory of fuzzy sets and of its extensions is in a way another testimonial to this technological queasiness.) An engineer, as opposed to a physicist, must present a client with a design that embodies certain performance assurances. To give these, at least ideally, he should utilize only facts and information that are known to him on a comparable confidence level. The trouble, of course, is that in practice much of his information is of lower quality and hence of much greater uncertainty. Chap. 1 exhibits a gallery of cases in point. These examples also show that probability theory cannot be relied upon for help. There are situations in which it merely compounds the difficulty, even when its application makes superficial sense. In others, it makes no sense in the first place.

The customary way out of the dilemma for the engineer has been his trusted safety factor. He has done so on the theory that it is better to over-design, and feel comfortable with one's performance assurances, than to under-design and run the risk of a failure. In other words, it is better to be wasteful than to court embarrassment or danger.

Professors Ben-Haim and Elishakoff have a better way. They point out in this book that the designer should make clear to himself what he knows on a sufficiently high confidence level, and then use this knowledge to place bounds on the design data and parameters that he can properly utilize. In effect, this will isolate for him a set of possible designs. The book deals with the case in which this set is convex and demonstrates that a very broad range of problems in applied mechanics actually satisfies the convexity requirement. Their approach is novel and highly welcome. In my opinion, it is inevitable that it, and its extensions, will dominate the future practice of engineering.

Professor-Emeritus Rudolf F. Drenick
Polytechnic Institute of New York
Huntington, New York
August, 1989

Contents

Preface

> If one thing today is certain, it is a feeling of uncertainty — a premonition that the future cannot be a simple extension of the present.[1]

Dear Reader:

The monograph before you deals with analysis of uncertainty in applied mechanics. It is an ambitious book. It is not just one more rendition of the traditional treatment of the subject, nor is it intended to supplement existing structural engineering books. The ultimate goal of this volume is to fill what the writers are confident is a significant gap in the available professional literature. This book is about *non-probabilistic* modelling of uncertainty. We have no ambition to supplant probabilistic ideas. Rather, we wish to provide an alternative avenue for analysis of uncertainty when only a limited amount of information is available. Whether this far-reaching goal has been successfully achieved is for you to judge.

Applied mechanics monographs and textbooks and associated university courses have historically focussed on the basic principles of mechanics, analysis, numerical evaluation and design. These books and courses almost exclusively assume that the material properties, geometric dimensions, and loading of the structure are precisely known.

Nobody questions the ever-present need for a solid foundation in applied mechanics. Neither does anyone question nowadays the fundamental necessity to recognize that uncertainty exists, to learn to evaluate it rationally, and to incorporate it into design.

But what do we mean by "uncertain"? Webster's Ninth New Colle-

[1] Queen Beatrix of the Netherlands, speech of allegiance to the Dutch constitution, April 30, 1980.

giate Dictionary provides the following definitions: "Indefinite, indeterminate; not certain to occur; problematical; not reliable; untrustworthy; not known beyond doubt; dubious; not having certain knowledge; doubtful; not clearly identified or defined; not constant; variable, fitful". Among the numerous synonyms to "uncertainty" one finds (Rodale, 1978): unsureness, unpredictability, randomness, haphazardness, arbitrariness, indefiniteness, indeterminacy, ambiguity, variability, changeability, irregularity, and so forth.

Recognition of the need to introduce the ideas of uncertainty in a wide variety of scientific fields today reflects in part some of the profound changes in science and engineering over the last decades.

The natural question arises: how to deal with uncertainty? Since one of the meanings of uncertainty is randomness, a natural answer to this question was and is to apply the theory of probability and random functions. Indeed, since the pioneering work by Meier in Germany, and by Freudenthal in Israel and later in the United States, probabilistic structural mechanics has achieved a high degree of sophistication. The power of probabilistic methods has been demonstrated beyond doubt in numerous publications on engineering subjects of paramount importance. Probabilistic modelling requires extensive knowledge of the random variables or functions involved. It leads to evaluation of the probability of successful performance by the structure, called reliability, or to its complement, the probability of failure. Is probabilistic modelling the only way one could deal with uncertainty? It turns out that the answer is negative. Indeed, the indeterminacy about the uncertain variables involved could be stated in terms of these variables belonging to certain sets, such as:

1. The uncertain parameter x is bounded, $|x| \leq a$.

2. The uncertain function has envelope bounds,

$$x_{\text{lower}}(t) \leq x(t) \leq x_{\text{upper}}(t)$$

   where $x_{\text{lower}}(t)$ and $x_{\text{upper}}(t)$ are deterministic functions which delimit the range of variation of $x(t)$.

3. The uncertain function has an integral square bound:

$$\int_{-\infty}^{\infty} x^2(t)\, dt \leq a$$

More examples will be found throughout the book. Instead of precise information on the probability content of random events, we possess imperfect, scanty knowledge on the uncertain quantities. This description of uncertainty is a set-theoretic, non-probabilistic one. We will limit ourselves, in this monograph, to representation of uncertain phenomena by *convex sets,* and will refer to this approach as *convex modelling.*

This set-theoretic approach was independently and almost simultaneously pioneered by Schweppe and by Drenick in 1968, the former in control theory and the latter in applied mechanics, namely in modelling of the response of structures to earthquake excitation. Schweppe summarized much of the early work in non-probabilistic modelling of uncertainty in his monograph *Uncertain Dynamic Systems* (1973). Independently, and without knowledge of these works, convex models of uncertainty were applied by Ben-Haim (1985) in his monograph on assay of material with uncertain spatial variation.

There is a similarity between probabilistic and convex modelling. Probabilistic modelling provides the probability of successful performance which is, roughly speaking, the probability of avoidance of the distribution tails of the random variables controlling the system. In similar fashion, convex modelling provides an evaluation based on assessments of extremal or delimiting properties of the system. In both circumstances, uncertain parameters lead to certain quantitative descriptions, but of different kinds.

Is there a contradiction between probabilistic and convex modelling?

We want to preface the answer to this question by more general considerations. Since the emergence of modern scientific methodology, to *understand* a natural phenomenon has meant to formulate a *conceptual model* which describes or predicts selected aspects of the phenomenon. To understand is to control, and the scientific model is the pre-eminent tool of the engineer in his quest for controlling man's environment. The scientific model is a distillation of reality. In attempting to explain this reality — to describe it qualitatively and quantitatively — we find that the same phenomenon can be represented by more than one model.

The validity of each model can be established by checking its compatibility with other models or with experimental observations. This done, the question naturally arises: which of these models is preferable? The answer to this question lies in adopting a criterion and setting an objective in constructing the model. The criterion in question is clearly one of usefulness.

Different models may refer to different aspects of a phenomenon and yield answers to specific questions in each case. The utilitarian approach to evaluating the multiplicity of models is both pragmatic and tolerant. Different models can "peacefully coexist" as long as they serve different purposes.

So the answer to our question as to possible contradiction between probabilistic and convex modelling is negative. The engineer is a most pragmatic creature and willingly allows coexistence of both models, as long as each fills a useful role. The engineer with professional responsibility, whose aim is in-depth study of complex phenomena, cannot allow himself to discard any useful model.

The organization of the monograph is as follows. The first chapter briefly reviews probabilistic methods and discusses the sensitivity of the probability of failure to uncertain knowledge of the system. Chapter two discusses the mathematical background of convex modelling. In the remainder of the book, convex modelling is applied to various linear and nonlinear problems. Except for the first two chapters, the material is almost exclusively based on our joint research, which continuously stimulates us to further the development of the subject, in nearly perfect cooperation and harmony (not free from an occasional "fight" on this or that idea, sentence or word). Accordingly, the sole responsibility for mistakes, misprints and misconceptions, as well as for successful achievement of our goals (which we hope for and count on) rests on both of us. One of our goals will be met if a serious reader, after initial perusal, is encouraged to read the book and feels that "it makes sense" despite the lack of esoteric mathematics.

Many disciplines originate in the efforts of a small group of researchers. When the ideas generated are seen to be related to real problems, they spread to a larger group. New questions then arise, stimulating the growth and spread of the discipline. If this happens as a result this book, then our goal will have been more than fully satisfied.

We wish to express our gratitude to the Technion — Israel Institute of Technology for providing an atmosphere which stimulates fruitful collaboration among its staff. Isaac Elishakoff appreciates the partial financial support of the Center of Applied Stochastics Research and the Department of Mechanical Engineering, Florida Atlantic University, during the final stages of this project. We are deeply indebted to Ms. Lilian Bluestein whose expert technical typing and unending patience have been invaluable in the preparation of the manuscript. We also wish to thank Mr. Eliezer Goldberg for valuable assistance in translating Russian texts, and Ms. Irit Nizan for her skillful technical drawings. We gratefully acknowledge the partial financial support provided by the Technion Fund for Promotion of Research, as well as a generous grant by the Technion - Niedersachsen (West Germany) Research Cooperation Fund.

Y. Ben-Haim, I. Elishakoff

Haifa and Boca Raton

Chapter 1

Probabilistic Modelling: Pros and Cons

> Idealists, materialists, realists, and empiricists have always had hopes of achieving some decisive formulation of their positions that would compell assent. But the differences in thought persist. Proofs are never so comprehensive as to preclude some revision or restatement of the position that was presumed destroyed. The pluralist abandons hope of any comprehensive resolutions of these conflicts and accepts them as distinct systems of thought with appropriate patterns of behavior and action.[1]

1.1 Preliminary Considerations

The countless achievements of traditional engineering mechanics are unimpeachable. Modern deterministic methods have become quite elaborate and include sophisticated mathematical modelling and analysis, highly refined methods of computer evaluation, and techniques of optimization. On the other hand as Alfred M. Freudenthal, widely recognized as a founder of modern stochastic structural mechanics, rightfully remarks,[2] "...it seems absurd to strive for more and more refinement of methods of stress-analysis if, in order to determine the dimen-

---

[1] Usher, 1954, p 28.

[2] Quoted from Ghiocel and Lungu, 1975, p 135.

sions of the structural elements, its results are subsequently compared with so-called working stress, derived in a rather crude manner by dividing the values of somewhat dubious material parameters obtained in conventional materials tests by still more dubious empirical numbers called safety factors." The rapid florescence of stochastic methods has indisputably enhanced the power of the designer: The concept of reliability has become a powerful tool which enables designers to deal probabilistically with uncertainty, when sufficient knowledge of the random variables and functions is available.

In order to place the study of convex models of uncertainty in its proper context, this chapter will begin by elucidating the basic facets of probabilistic design. Following this is an examination of the difficulties which arise when either the underlying deterministic model is not perfectly chosen, or is simply unavailable, or when there is incomplete knowledge of the probabilistic variables involved. For better grasp by the reader (who need not be a specialist in stochastic mechanics) this discussion is accompanied by a multiplicity of simple illustrative examples. At the end of the chapter we present a podium for debate through extensive yet instructive quotations on the philosophy of probabilistic design.

1.2 Probabilistic Modelling in Mechanics

1.2.1 Reliability of Structures

A structure can be modelled as a system subjected to influences, disturbances, excitations, and so forth, designated here as inputs. We are usually interested in the response of this sytem — its displacements, stresses, strains, and so on, designated as outputs. In formulating a probabilistic model we have first to identify the random variables, which can be the characteristics of the inputs and/or of the system itself. Once the random variables and functions are identified, we need to know their joint probability density function. The beauty of probabilistic modelling lies in the fact that it can be directly incorporated in design. Within the probabilistic context the main objective of en-

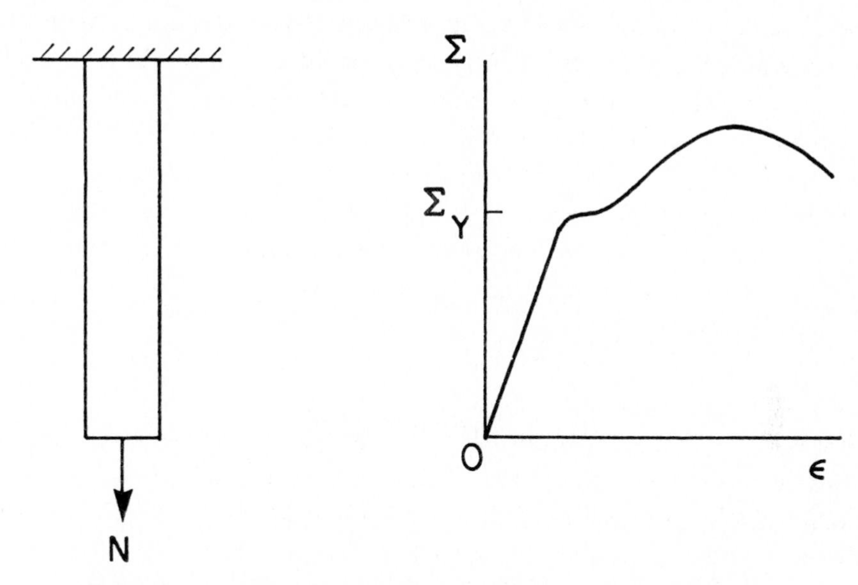

Figure 1.1: Tension member with exponential loading.

gineering analysis is the determination of the *reliability* of the system. Reliability is defined as the probability that the system will adequately perform its duty for a specified interval of time, when operating under stated environmental conditions.

This notion of reliability admits the possibility of failure. It is usually exceedingly difficult to define failure in unambiguous terms. Of course, complete and catastrophic failure is easily recognized. On the other hand, system performance can deteriorate gradually over time. Once the function and failure modes of the system are explicitly stated, reliability can be quantified with precision by probabilistic calculations. We will illustrate this with several examples.

Example. Tension Member with Exponential Loading. The reliability of a tension member may be defined as the probability of the actual stress being lower than the yield stress (Fig. 1.1). If the cross-sectional area of the bar is a deterministic quantity but the applied load N and yield stress Σ_y are random variables with given probability densities $P_N(n)$ and $P_{\Sigma_y}(\sigma_y)$ respectively and joint probability densities $P_{N\Sigma_y}(n, \sigma_y)$ then the equilibrium equation of the bar is:

$$b\Sigma = N \tag{1.1}$$

where Σ is the actual stress in its cross section and b is the cross-section area. The reliability R can then be defined as

$$R = \text{Prob}(\Sigma \le \Sigma_y) \tag{1.2}$$

or

$$R = \text{Prob}(\Sigma = N/b \le \Sigma_y) \tag{1.3}$$

Alternatively, knowing the joint probability density $P_{N\Sigma_y}(n, \sigma_y)$, we can define the reliability in integral form

$$R = \int_A P_{N\Sigma_y}(n, \sigma_y) dn d\sigma_y \tag{1.4}$$

where the integration interval is as per Fig. 1.2. Then

$$R = \int_0^\infty \int_{n/b}^\infty P_{N\Sigma_y}(n, \sigma_y) dn d\sigma_y \tag{1.5}$$

Let N and Σ_y be independent and exponentially distributed, with probability densities

$$p_N(n) = \begin{cases} \alpha_1 e^{-\alpha_1 n} & n > 0 \\ 0 & \text{otherwise} \end{cases} \tag{1.6}$$

and

$$p_{\Sigma_y} = \begin{cases} \alpha_2 e^{-\alpha_2 \sigma_y} & \sigma_y > 0 \\ 0 & \text{otherwise} \end{cases} \tag{1.7}$$

Then

$$\begin{aligned} R &= \int_0^\infty dn \int_{n/b}^\infty \alpha_1 \alpha_2 e^{-\alpha_1 n - \alpha_2 \sigma_y} d\sigma_y & (1.8) \\ &= \int_0^\infty \alpha_1 \alpha_2 e^{-\alpha_1 n} \left(\frac{e^{-\alpha_2 n/b}}{\alpha_2} \right) dn & (1.9) \\ &= \frac{\alpha_1}{\alpha_1 + \alpha_2/b} & (1.10) \end{aligned}$$

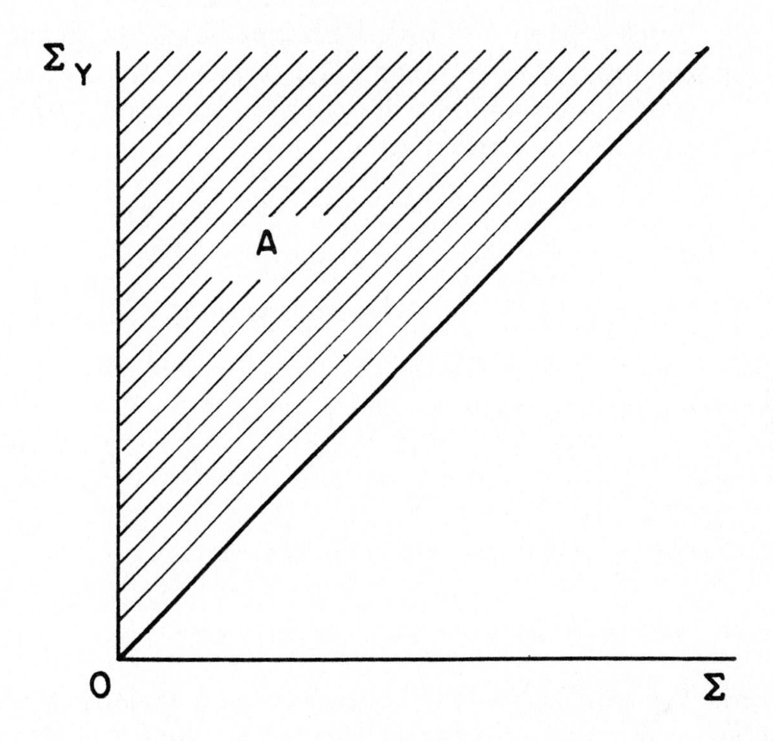

Figure 1.2: Domain of integration for evaluating the reliability.

With this expression available, we are able to solve a design problem. The design requirement reads then

$$R \geq r \tag{1.11}$$

where r is the reliability specified for an ensemble of macroscopically identical structures; r is a codified quantity and reflects, among other things, the degree to which successful performance of the structure is important to the user. Combining eqs.(1.10) and (1.11) the design requirement takes the form:

$$\frac{\alpha_1}{\alpha_1 + \alpha_2/b} \geq r \tag{1.12}$$

With α_1 and α_2 given, this inequality determines the minimum required area of the member

$$b \geq b_{\text{min,req}} = \frac{r}{1-r}\frac{\alpha_2}{\alpha_1} \tag{1.13}$$

Example. Tension Member with Uniform Loading. Consider the tension member referred to earlier (Fig. 1.1), subjected to a random loading with a uniform probability density $p_N(n)$ on the interval $[n_1, n_2]$. The cumulative probability distribution is:

$$F_N(n) = \begin{cases} 0 & \text{for} \quad n < n \\ \dfrac{n - n_1}{n_2 - n_1} & \text{for} \quad n_1 \leq n \leq n_2 \\ 1 & \text{for} \quad n > n_2 \end{cases} \tag{1.14}$$

We will consider the case when the yield stress is a deterministic quantity, σ_y. The actual stress is again a random variable

$$\Sigma = \frac{N}{b} \tag{1.15}$$

where b is the cross-sectional area. The reliability is then

$$R = \text{Prob}\left(\Sigma = \frac{N}{b} < \sigma_y\right) = F_N(b\sigma_y) \tag{1.16}$$

In other words, the reliability of the bar equals the cumulative distribution function of the tensile load at the $b\sigma_y$ level. Thus

$$R = \begin{cases} 0 & \text{for} \quad b\sigma_y < n \\ \dfrac{b\sigma_y - n_1}{n_2 - n_1} & \text{for} \quad n_1 \leq b\sigma_y \leq n_2 \\ 1 & \text{for} \quad b\sigma_y > n_2 \end{cases} \tag{1.17}$$

The reliability requirement

$$\frac{b\sigma_y - n_1}{n_2 - n_1} \geq r \tag{1.18}$$

where r is the required reliability, yields the minimum required cross-sectional area:

$$b_{\text{min,req}} = \frac{r(n_2 - n_1) + n_1}{\sigma_y} \tag{1.19}$$

We can express $b_{\text{min,req}}$ in terms of the mean $E(N)$ and the standard deviation σ_N of the load

$$E(N) = \frac{1}{2}(n_1 + n_2) \quad , \quad \sigma_N = \frac{1}{\sqrt{12}}(n_2 - n_1) \tag{1.20}$$

to result in

$$b_{\mathrm{min,req}} = \frac{E(N) + (r - 0.5)\sqrt{12}\sigma_N}{\sigma_y} \tag{1.21}$$

This formula indicates that in this case the probabilistic design includes the perfectly deterministic one as a particular case. Indeed, if $n_2 \to n_1$, then $E(N) \to n_1$ and $\sigma_N \to 0$, and $b_{\mathrm{min,req}}$ tends to $E(N)/\sigma_y = n_1/\sigma_y$ which is obtained through the deterministic design.

An analogous formula, with similar conclusions, is obtained for normally distributed N. Here Φ is an integral distribution function of the standard normal variable, and the required minimum value of b is:

$$b_{\mathrm{min,req}} = \frac{E(N) + \sigma_N \Phi^{-1}(r)}{\sigma_y} \tag{1.22}$$

Example. Buckling of a Thin Shell. Buckling has been an all-important topic[3] since Euler in 1744 formulated his classical equation for the buckling load of a bar of length L, made of a material with elastic modulus E and cross-sectional moment of inertia I, and simply supported at both ends (Fig. 1.3):

$$P_{cl} = \frac{\pi^2 EI}{L^2} \tag{1.23}$$

Since Euler's pioneering work, many thousands of investigations have been conducted on buckling of bars, plates and shells. Inter alia Euler's formula was generalized in this century for circular cylindrical shells. The expression refers to the classical buckling stress

$$\sigma_{cl} = \frac{1}{\sqrt{3(1-\nu^2)}} E \frac{h}{R} \tag{1.24}$$

where ν is Poissons ratio, h the thickness of the shell and R its radius. The classical buckling load is $P_{cl} = 2\pi R h \sigma_{cl}$. It turned out, however, that experimental buckling loads exhibit wide scatter, and moreover are much lower than those predicted by the classical theory of elastic stability. This discrepancy attracted many investigators. Koiter in

[3]Some specialists maintain that the first recorded buckling failure was that of the Tower of Babel, which must have collapsed under its own weight.

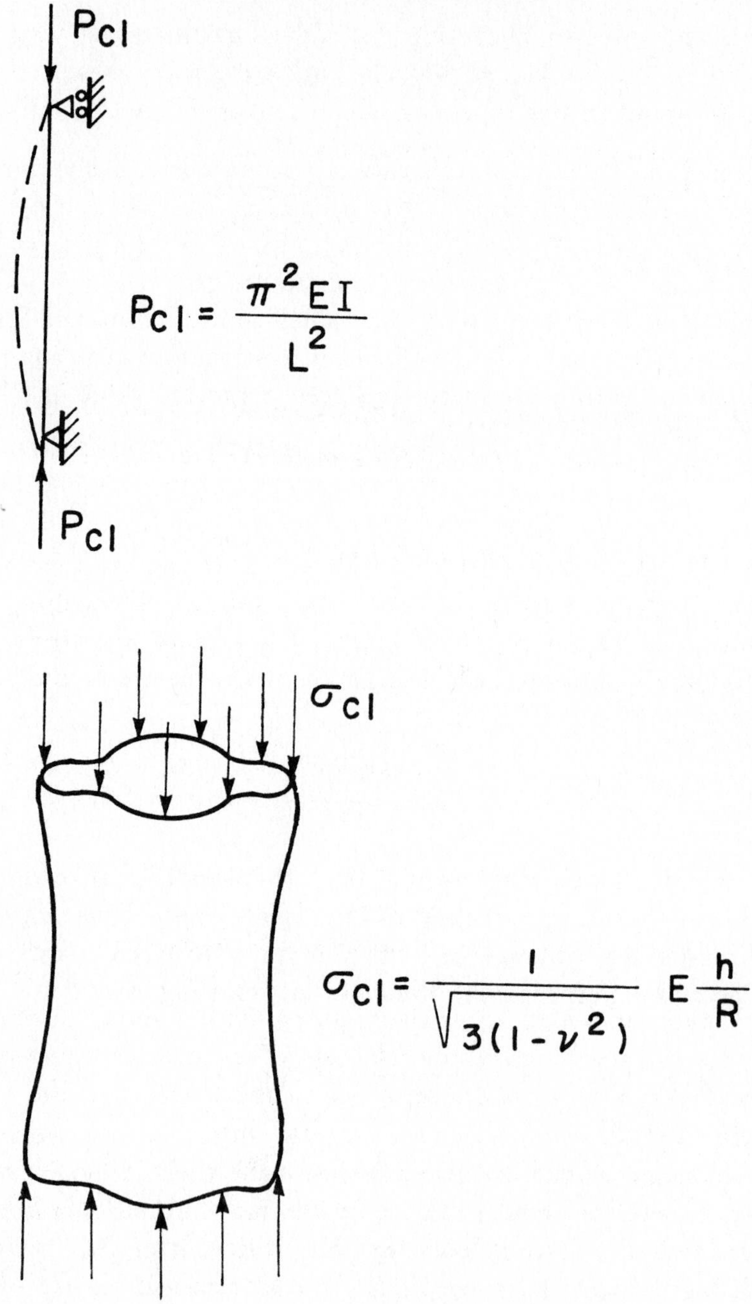

$$P_{cl} = \frac{\pi^2 EI}{L^2}$$

$$\sigma_{cl} = \frac{1}{\sqrt{3(1-\nu^2)}} E \frac{h}{R}$$

Figure 1.3: Buckling of a thin shell.

1945, almost exactly two centuries after Euler, made a most important contribution to the buckling literature in his doctoral thesis. He discovered that small initial imperfections — geometric deviations from nominal ideal cylindricality — may cause significant reduction of the load-carrying capacity of the shell. He represented an initial imperfection function $w_o(x)$ as the axisymmetric buckling mode of a nominally perfect cylindrical shell, namely

$$w_o(x) = \mu h \sin \frac{i\pi x}{L} \tag{1.25}$$

where μ is the (nondimensional) initial imperfection amplitude, i is the number of half-waves at which the associated perfect shell would buckle

$$i = \frac{L}{\pi}\sqrt{\frac{2c}{Rh}} \quad , \quad c = \sqrt{3(1-\nu^2)} \tag{1.26}$$

L shell length, R the shell radius, h the shell thickness. Via his general nonlinear theory, Koiter derived a relationship between the critical load and the initial imperfection amplitude:

$$(1-\lambda)^2 - \frac{3}{2}c|\mu|\lambda = 0 \tag{1.27}$$

where

$$\lambda = \frac{P_{bif}}{P_{cl}} \tag{1.28}$$

P_{bif} is the buckling load of an imperfect shell and P_{cl} — that of the associated perfect shell, given in eq. (1.24). The buckling load P_{bif} was defined as that at which the axisymmetric fundamental equilibrium state bifurcates into a nonsymmetric one. The absolute value of μ is used in eq.(1.27) as for a sufficiently long shell the sign of the imperfection is immaterial: Positive and negative initial imperfections with the same absolute value cause the same reduction of the buckling load.

Equation (1.27) shows that its solution for λ should be less than unity, since $|\mu| \ll 1$. This implies that the critical buckling load of the imperfection structure is lower than its perfect counterpart. Of course, if we put $\mu \equiv 0$, we have $\lambda = 1$ or $P_{bif} = P_{cl}$. In other words, the above decrease in the critical buckling load is due to presence of the (unavoidable) initial imperfections.

The formula (1.27) allows an immediate probabilistic generalization. It shows us how to find the probabilistic characteristics of the random buckling load P_{bif} (with possible values of P_{bif}) provided the critical imperfection μ (with possible values of μ) is a random variable with given probability distribution $F_\mu(\mu)$. For us, as before, what mainly counts is the reliability of the structure, which now is defined as the probability of the random event that the load P_{bif}/P_{cl} exceeds a given nondimensional load α

$$R(\alpha) = \text{Prob}\left(\frac{P_{bif}}{P_{cl}} > \alpha\right) \tag{1.29}$$

i.e., the structure does not buckle prior to a specified load α, or in other words "lives" above α. From eq.(1.27) it follows that P_{bif}/p_{cl} exceeds α if

$$|\mu| < \frac{2(1-\alpha)^2}{3c\alpha} \tag{1.30}$$

Then

$$R(\alpha) = \text{Prob}\left(|\mu| < \frac{2(1-\alpha)^2}{3c\alpha}\right) = F_{|\mu|}\left(\frac{2(1-\alpha)^2}{3c\alpha}\right) \tag{1.31}$$

Then

$$R(\alpha) = \text{Prob}\left(-\frac{2(1-\alpha)^2}{3c\alpha} < \mu < \frac{2(1-\alpha)^2}{3c\alpha}\right) \tag{1.32}$$

$$= \Phi\left(\frac{2(1-\alpha)^2}{3c\alpha\sigma} + \frac{a_\mu}{\sigma}\right) - \Phi\left(-\frac{2(1-\alpha)^2}{3c\alpha\sigma} - \frac{a_r}{\sigma}\right) \tag{1.33}$$

This means that once the probability distribution function of the initial imperfections is known, one can proceed with the determination of the reliability $R(\alpha)$, or of its compliment, probability of failure $P_f = 1 - R(\alpha)$ (see Bolotin, 1969; Roorda, 1980; and Elishakoff, 1983). Here also, as in the case of the design for strength or stiffness, the probabilistic calculations are sensitive to both the deterministic theory used and to the imperfect information on the probabilistic characteristics. This again is better considered by simple examples.

1.2.2 Sensitivity of Failure Probability

Let us consider some examples which are elementary in themselves, but in which the reliability calculations are highly sensitive to the theory used. Since the strength of materials formulas ordinarily used are approximate, the predictions they yield may deviate widely from the actual reliability calculated with an exact theory. A few straightforward examples follow.

Example. Tube with Internal Pressure. Consider a tube loaded by an internal pressure P (Fig. 1.4). The circumferential and radial stresses are determined, respectively, by the following theory-of-elasticity (exact) expressions

$$\sigma_\theta\big|_{r=a} = P\frac{b^2+a^2}{b^2-a^2}\ , \quad \sigma_r\big|_{r=a} = -p \tag{1.34}$$

with the appropriate equivalent stress within the maximum shear stress criterion according to Tresca:

$$\sigma_{\text{eq}} = \sigma_1 - \sigma_3 = \sigma_\theta - \sigma_r = P\frac{2b^2}{b^2-a^2} \tag{1.35}$$

Now, if the outer radius is very close to the inner one, i.e. if the wall thickness of the tube is much smaller than its inner radius

$$h = b - a \ll a \tag{1.36}$$

Then the following approximate formula holds for the circumferential stress at $r = a$:

$$\sigma_\theta\big|_{r=a} = P\frac{a}{h} \tag{1.37}$$

Let now the inner pressure be distributed exponentially with mean $E(P)$ and let the acceptability criterion set the codified probability of failure at $P_f^* = 10^{-7}$.

The probability distribution of P is

$$F_P(p) = 1 - \mathrm{e}^{-p/E(P)} \tag{1.38}$$

and within the approximate strength-of-materials theory,

$$\sigma_{\text{eq}} = P\frac{a}{h} - (-P) = P\frac{b}{h}. \tag{1.39}$$

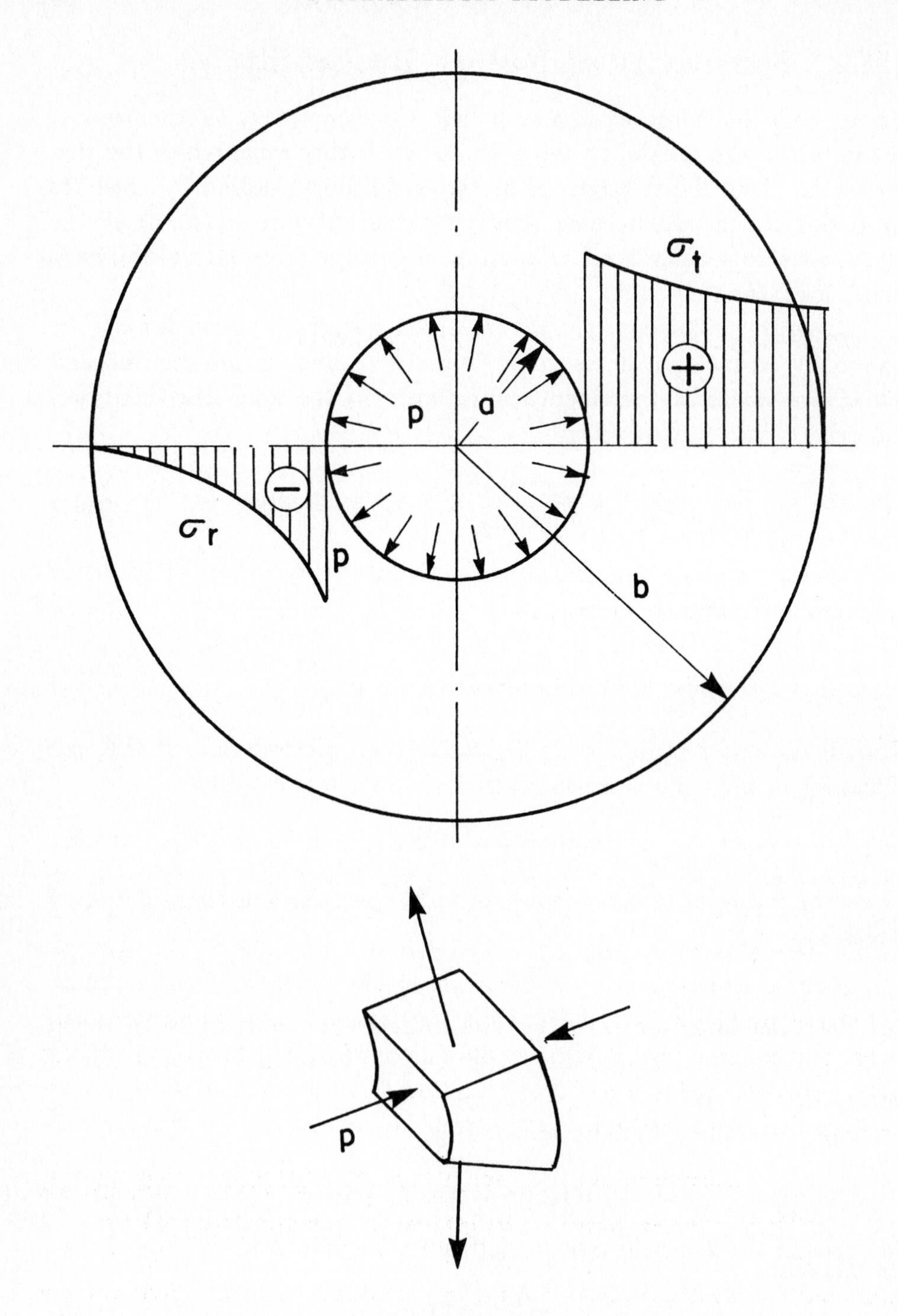

Figure 1.4: Tube with internal pressure.

Now, σ_Y being the yield stress, the estimate of the probability of failure is:

$$P_f = \text{Prob}(\sigma_{eq} > \sigma_Y) = 1 - \text{Prob}\left(P \leq \frac{\sigma_Y h}{b}\right) \tag{1.40}$$

$$= \exp\left(-\frac{\sigma_Y h}{bE(P)}\right) \tag{1.41}$$

We calculate $E(P)$ so that the probability of failure equals P_f^*

$$E(P) = -\frac{h\sigma_Y}{b\ln P_f^*} \tag{1.42}$$

We want to know to what actual probability of failure P_f this $E(P)$ corresponds within the exact theory:

$$P_f = \text{Prob}\left(p\frac{2b^2}{b^2 - a^2} > \sigma_Y\right) = \exp\left(-\frac{\sigma_Y(b^2 - a^2)}{2b^2 E(P)}\right) \tag{1.43}$$

or, with $E(P)$ from eq.(1.42), we finally obtain

$$P_f = \exp\left(\frac{b+a}{2b}\ln P_f^*\right) = \left(P_f^*\right)^{(b+a)/2b} \tag{1.44}$$

Let $a/b = 0.97$, which corresponds to a 1.5% error in the circumferential stress at the inner boundary — quite acceptable by engineering standards. Then, with $P_f^* = 10^{-7}$, we find $P_f = 10^{-6.895}$, i.e. the actual failure probability exceeds the codified one. For $a/b = 0.91$ (corresponding to less than five percent error in calculation of circumferential stress in eq.(1.37)), the actual failure probability exceeds more than twice the codified value. This implies that with $E(P)$ as in eq.(1.42), the designer would accept the element on the basis of the approximate analysis, whereas the exact result would contra-indicate such acceptance.

Example. Beam Under Sudden Loading. Consider a beam, simply supported at its ends, subjected to a suddenly applied load at its midspan:

$$q(x,t) = P\delta\left(x - \frac{\ell}{2}\right) \tag{1.45}$$

where P is the magnitude of the force, ℓ is the length of the beam, $\delta(\cdot)$ is a Dirac delta function. Elementary solution for the displacement response of the beam yields

$$w(x,t) = \frac{2P}{\ell}\sum_{k=1}^{\infty} \sin\frac{k\pi}{2}\frac{1-\cos\omega_k t}{\omega_k^2}\sin\frac{k\pi x}{\ell} \tag{1.46}$$

where ω_k is the kth natural frequency, k is the serial number of the mode shape, number of half-waves of the mode shape in the longitudinal vibration. For the midspan we arrive at

$$w(x,t) = \frac{2P}{\ell}\sum_{k=1,3,5,\ldots}\frac{1-\cos\omega_k t}{\omega_k^2} \tag{1.47}$$

Within the classical Bernoulli-Euler theory, which neglects the rotary inertia and shear deformation,

$$\omega_k^2 = \left(\frac{k\pi}{\ell}\right)^4\frac{EI}{\rho A} \tag{1.48}$$

where E is the modulus of elasticity, I is the moment of inertia, ρ is the density and A is the cross sectional area.

The series in eq.(1.47) is rapidly convergent, so one can approximate the system as one possessing a single degree of freedom:

$$w\left(\frac{\ell}{2},t\right) = \frac{2P}{\rho A\ell}\frac{1-\cos\omega_1 t}{\omega_1^2} \tag{1.49}$$

We assume now that P is a random variable and that acceptability of a structure is associated with fulfillment of the inequality $w_{max} < w^*$, where ω^* is a codified displacement. The probability of failure is then

$$P_f = \text{Prob}(w_{max} > w^*) \tag{1.50}$$

Now

$$w_{max} = w\left(\frac{\ell}{2},\frac{\pi}{\omega_1}\right) = \frac{4P}{\rho A\ell}\frac{1}{\omega_1^2} \tag{1.51}$$

For the simplicity of calculations we will assume that P is an exponentially distributed random variable

$$F_P(p) = 1-\exp\left(-\frac{p}{E(P)}\right) \tag{1.52}$$

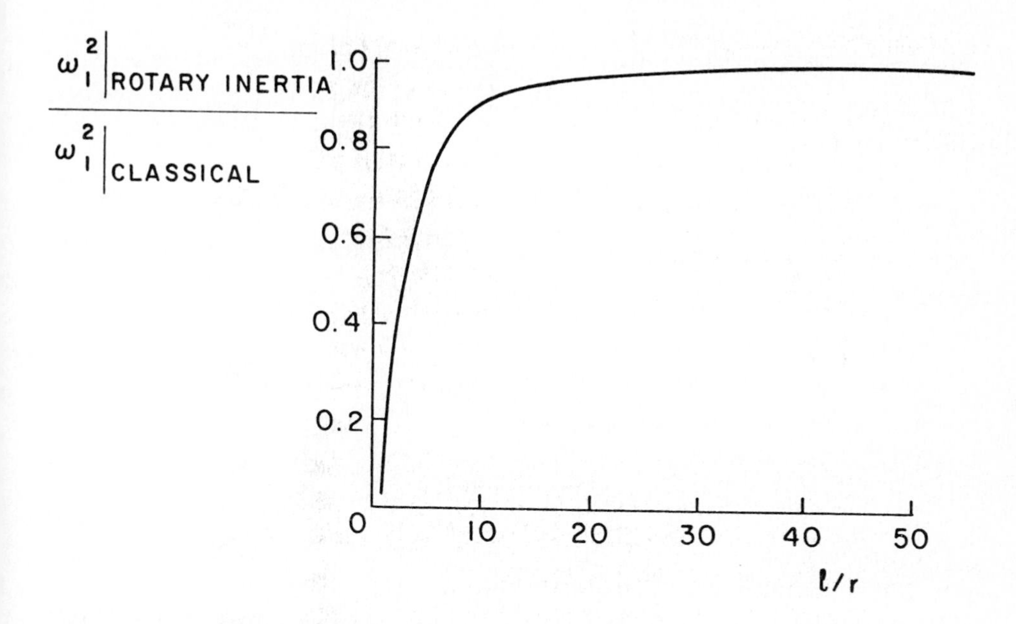

Figure 1.5: Ratio of natural frequency with rotary inertia to the classical natural frequency.

We will set $E(P)$ so that the probability of failure equals its codified value P_f^* :

$$\mathrm{Prob}(w_{\max} > w^*) = P_f^* \tag{1.53}$$

Then

$$\exp\left(-\frac{w^* \rho A \ell \omega_1^2}{4E(P)}\right) = P_f^* \tag{1.54}$$

and

$$E(P) = -\frac{w^* \rho A \ell \omega_1^2}{4 \ln P_f^*} \tag{1.55}$$

The refinement of the classical Bernoulli-Euler theory, which takes into account the rotary inertia but still neglects the shear deformation yields the following estimate of the natural frequency (Timoshenko, 1928):

$$\omega_1^2\Big|_{\text{rotary inertia}} = \frac{\omega_{1,\text{class}}^2}{1 + (\pi r/\ell)^2} \tag{1.56}$$

where $r = \sqrt{i/a}$ is the radius of gyration of the cross section (Fig. 1.5). Under these circumstances the probability of failure, within this refined theory becomes

$$P_f\big|_{\text{rotary inertia}} = \text{Prob}\left(w_{\max}\big|_{\text{rotary inertia}} > w^*\right) \tag{1.57}$$

where in the expression for the maximum displacement one should substitute a refined value of the natural frequency. Hence

$$P_f\big|_{\text{rotary inertia}} = \exp\left(-\frac{w^*\rho A\ell}{4E(P)}\,\omega_1^2\big|_{\text{rotary inertia}}\right) \tag{1.58}$$

$$= \exp\left(-\frac{w^*\rho A\ell\omega_1^2}{4E(P)[1+(\pi r/\ell)^2]}\right) \tag{1.59}$$

$$= (P_f^*)^{[1+(\pi r/\ell)^2]^{-1}} \tag{1.60}$$

For $\ell/r = 15$, and $P_f^* = 10^{-7}$, we get the estimate of the actual probability of failure with rotary inertia taken into account to be about 2×10^{-7}, twice the codified value; for $\ell/r = 10$, the refinement of the probability of failure is more than four times the codified value.

Now another approximation, which takes into account the shear deformation, leads to

$$\omega_1^2\big|_{\text{shear}} = \frac{\omega_{1,\text{class}}^2}{1+(E/kG)(\pi r/\ell)^2} \tag{1.61}$$

where G is the shear modulus and k is the shear coefficient depending on the form of the cross-section (Fig. 1.6). This yields then a new estimate for the probability of failure

$$P_f\big|_{\text{shear}} = \left(P_f^*\right)^{[1+(E/kG)(\pi r/\ell)^2]^{-1}} \tag{1.62}$$

With $r/\ell = 15$, $E/kG = 3.06$ and $P_f^* = 10^{-7}$, we get the estimate of the actual probability of failure with shear deformation taken into account to be 6.74×10^{-7}, more than six times the codified value; for $\ell/r = 10$ the refinement yields 4.2×10^{-6}.

The "best" deterministic theory is the one due to Timoshenko (1928), which takes into account both the rotary inertia and shear deformation:

$$\omega_1^2\big|_{\text{Timoshenko}} \approx \frac{\omega_1^2}{1+(1+E/kG)(\pi r/\ell)^2} \tag{1.63}$$

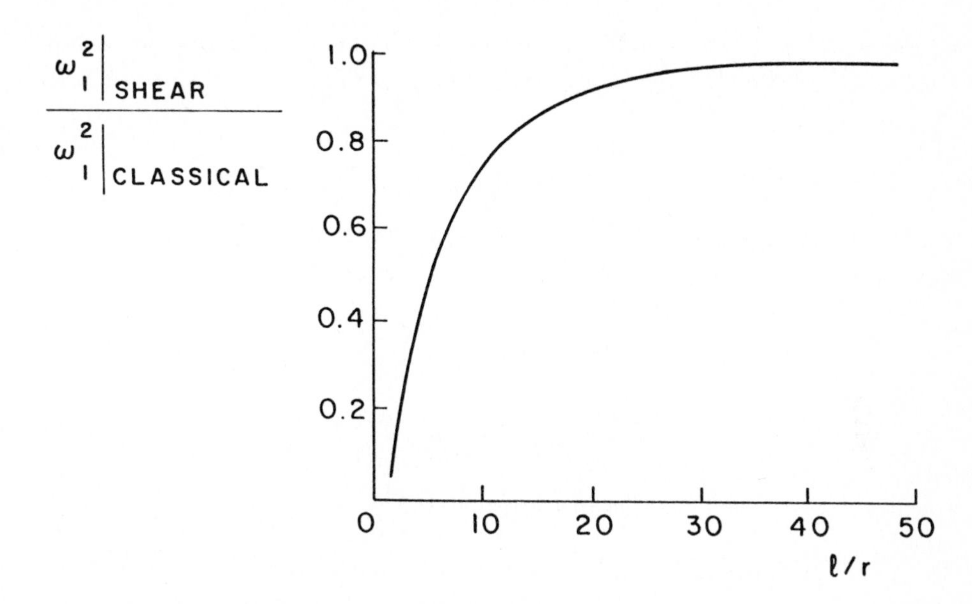

Figure 1.6: Ratio of natural frequency with shear to the classical natural frequency.

(Fig. 1.7), and, consequently,

$$P_f\big|_{\text{Timoshenko}} = \left(P_f^*\right)^{[1+(1+E/kG)(\pi r/\ell)^2]^{-1}} \tag{1.64}$$

Again for $\ell/r = 15$, $E/kG = 3.06$ and $P_f^* = 10^{-7}$, we get now that the actual probability of failure is 1.14×10^{-6} which is an order of magnitude more than the designed value. For the transversely isotropic beam, the ratio E/kG may assume considerable values; for say $E/kG = 10$, and the rest of data as above, we obtain $P_f \approx 2\times10^{-5}$ which is two orders of magnitudes more than the codified value.

As we have seen in the above examples, the application of the approximate theory, even when it yields only very few percentages of error within the deterministic calculation, may produce unsatisfactory results for the probability of failure. This is not always the case. We will show an example of the opposite, "positive" nature.

Example. Circular Clamped Plate. Consider a clamped plate under the action of an uniform transverse load $\bar{Q}_0$, which is a random variable with given distribution. The load–displacement relation

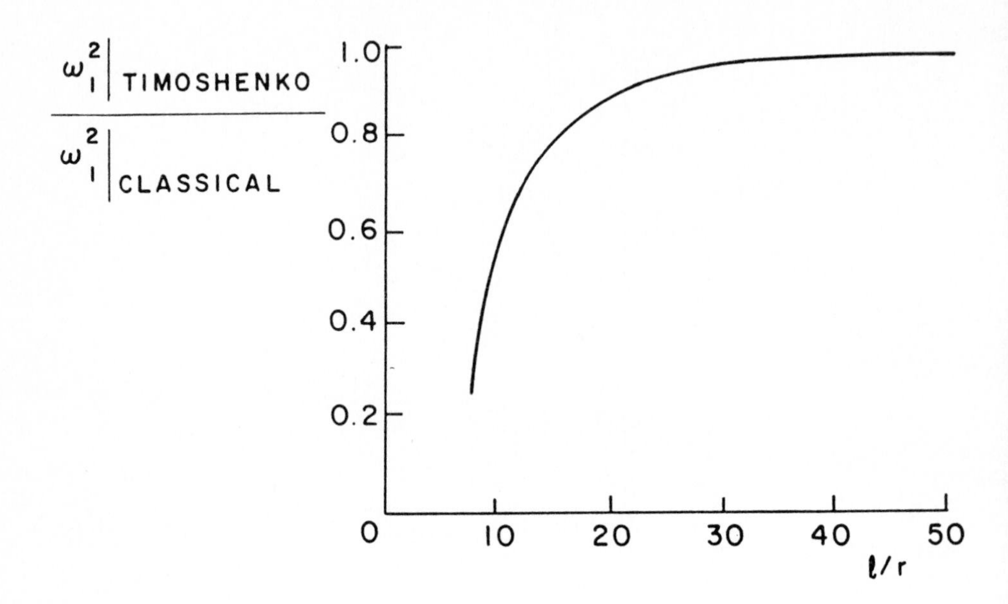

Figure 1.7: Ratio of natural frequency with rotary inertia and shear to the classical natural frequency.

is given by the relation (Dym, 1973):

$$\frac{W_0}{h} + 0.346\left(\frac{W_0}{h}\right)^3 = \bar{Q} \tag{1.65}$$

where W_0 is the maximum deflection occurring at the origin, h is the plate thickness, $\bar{Q}$ is the nondimensional loading

$$\bar{Q} = \frac{Q_0 a^4}{64hD} \tag{1.66}$$

where Q_0 is the actual loading, a is the plate radius, and D is the cylindrical stiffness $= Eh^3/12(1-\nu^2)$, E is modulus of elasticity and $\nu =$ Poisson's ratio. Fig. 1.8 shows a nonlinear load-deflection relationship. Now, if the possible values of $\bar{Q}$ are much less than unity, then the possible values of the displacements w_0 also turn out to be much less than the thickness h, and the second term in eq.(1.65) can be neglected. This implies that this relationship could then be linearized. The safe

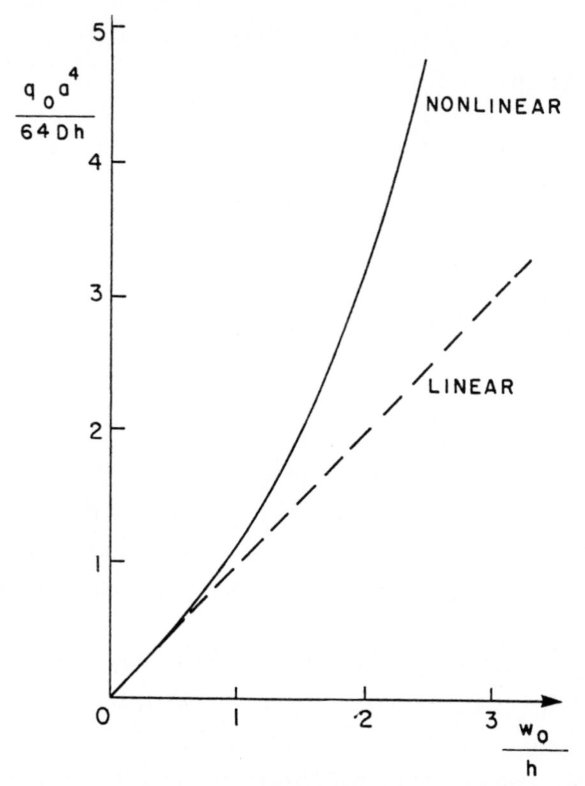

Figure 1.8: Linear and non-linear load-deflection relations for a clamped plate.

operation of the system will be identified with the maximum displacement being less than some fraction of the plate thickness αh ($\alpha < 1$). Within the linearized theory then

$$\mathrm{Prob}(W_0 \geq \alpha h) = \mathrm{Prob}(\bar{Q} \geq \alpha) \tag{1.67}$$

If we again adopt the exponential distribution for $\bar{Q}$ with the mean loading equal $E(\bar{Q})$ we have

$$\mathrm{Prob}(W_0 \geq \alpha h) = \exp\left(-\frac{\alpha}{E(\bar{Q})}\right) \tag{1.68}$$

We set $E(\bar{Q})$ so as to have a codified value of the probability of failure P_f^*. Hence

$$E(\bar{Q}) = -\frac{\alpha}{\ln P_f^*} \tag{1.69}$$

Now, within the more elaborate nonlinear theory the actual probability of failure turns out to be

$$P_f = \text{Prob}(W_0 \geq \alpha h) = \text{Prob}(\bar{Q} \geq \alpha + 0.346\alpha^3) \quad (1.70)$$

$$= \exp\left(-\frac{\alpha + 0.346\alpha^3}{E(Q)}\right) \quad (1.71)$$

With eq. (1.69) for $E(\bar{Q})$ we get finally

$$P_f = \left(P_f^*\right)^{1+0.346\alpha^2} \quad (1.72)$$

Since the power is more than unity and $P_f^* \ll 1$, we arrive at the conclusion that the actual probability of failure turns out to be smaller than the codified one. We should have expected a result of this nature, since for the hardening nonlinearity the calculation according to the linear theory should go to the safe side.

From these three examples we learn that using approximate deterministic theories may yield inaccurate estimates of the actual probability of failure. This is especially important if one keeps in mind that exact estimates of the probability of failure are known only in the simplest cases.

Example. Uncertain Exponential Tails. Consider now how imperfect information on the probability density will affect the structural design. Assume that the probability density of the loading is given by

$$f_P(p) = \begin{cases} \alpha\lambda_1 e^{-\lambda_1 p} & , \quad 0 \leq p \leq P_1 \\ \beta\lambda_2 e^{-\lambda_2 p} & , \quad p > P_1 \end{cases} \quad (1.73)$$

Due to continuity of the probability density at $p = P_1$, we have

$$\alpha\lambda_1 e^{-\lambda_1 P_1} = \beta\lambda_2 e^{-\lambda_2 P_1} \quad (1.74)$$

The normalization requirement yields

$$\int_0^{P_1} \alpha\lambda_1 e^{-\lambda_1 P}\, dp + \int_{P_1}^{\infty} \beta\lambda_2 e^{-\lambda_2 P}\, dp = 1 \quad (1.75)$$

The two latter equations allow one to find α and β in terms of specified λ_1, λ_2, and P_1 :

$$\alpha = \frac{1}{1 - e^{-\lambda_1 P_1}(1 - \lambda_1/\lambda_2)} \tag{1.76}$$

$$\beta = \alpha \frac{\lambda_1}{\lambda_2} \frac{e^{-\lambda_1 P_1}}{e^{-\lambda_2 P_1}} \tag{1.77}$$

If $\lambda_1 = \lambda_2$, we find $\beta = \alpha$, as expected. The more realistic situation $\lambda_1 \neq \lambda_2$ may describe the case when the exact density is given by eq.(1.73) but a judgement was made that one has here a conventional exponential density. This may especially be a realistic case since for large values of P the experimental values are difficult to obtain and one could be unaware of another exponential governing the behaviour at the tail.

The physical problem is a simple case of a tension bar with the yield stress being a deterministic quantity σ_Y. The probability of failure is set at P_f^*. Then if one made an imprecise judgement that one has a single exponential distribution with parameter λ_1, one gets for the failure probability

$$\text{Prob}(P \geq a\sigma_Y) = e^{-\lambda_1 a \sigma_Y} \tag{1.78}$$

where a is the cross-sectional area of the bar. If one faces the design problem, i.e. choosing a, one obtains

$$a = -\frac{\ln P_f^*}{\lambda_1 \sigma_Y} \tag{1.79}$$

Now in actuality

$$P_f = \text{Prob}(P \geq a\sigma_Y) = \int_{a\sigma_Y}^{\infty} f_P(p)\, dp \tag{1.80}$$

where f_P is given by eq. (1.73). If $a\sigma_Y \geq P_1$ we have

$$P_f = \int_{a\sigma_Y}^{\infty} \beta \lambda_2 e^{-\lambda_2 P}\, dp = \beta e^{-\lambda_2 a \sigma_Y} \tag{1.81}$$

With the above value of a we find an actual probability of failure as:

$$P_f = \beta \left(P_f^*\right)^{\lambda_2/\lambda_1} \tag{1.82}$$

Let us take the values $\lambda_1 = 1$, $\lambda_2 = \frac{1}{2}$, $\sigma_Y = 100\text{kg/cm}^2$, $P_f^* = 10^{-7}$. Then

$$\begin{aligned} a &= 0.0016118 && (1.83)\\ \alpha &= 0.9525741 && (1.84)\\ \beta &= 0.4254096 && (1.85)\\ P_f &= 1.34526 \times 10^{-4} && (1.86) \end{aligned}$$

which is three magnitudes more than the codified failure probability. Indeed, this failure probability is determined by the behaviour of the distribution at the tail, but the experimental information on the tail may be very imprecise.

Example. Normal Distribution with Uncertain Tails.[4] In this example, the high sensitivity (ill-conditioning) to the assumption of Gaussianity will be demonstrated, i.e. it will be demonstrated that the failure probability is most sensitive to those characteristics of the underlying random process $y(t)$ that are least likely to be well known, namely the behaviour at large $y(t)$. The probability of failure when $y(t)$ is a stationary Gaussian random process is given by Pickands (1969) in a closed form:

$$P(\max_t y(t) > L) = 1 - \exp\left\{-\exp\left[-\frac{L}{\sigma}(2\log 2n)^{\frac{1}{2}} + \eta\right]\right\} \qquad (1.87)$$

where n and η are constants. The first is a coefficient in the Maclaurin series for the autocovariance $R_y(\tau)$ of $y(t)$

$$R_y(\tau) = R_y(0)\left[1 - n\left|\frac{\tau}{T}\right| + O\left(\frac{\tau}{T}\right)\right] \qquad (1.88)$$

and the second is

$$\eta = 2\log(2n) + \frac{1}{2}(\log \pi - \log\log 2n) \qquad (1.89)$$

Eqs.(1.88) and (1.89) are asymptotically valid for large L and large n, in the sense that terms of order $O(L/\sigma)^{-1}$ and $O(\log n)^{-1}$ are neglected. Drenick (1977) assumed that there is an equivalence between

[4]This example is taken almost verbatim from Drenick (1977, 1979).

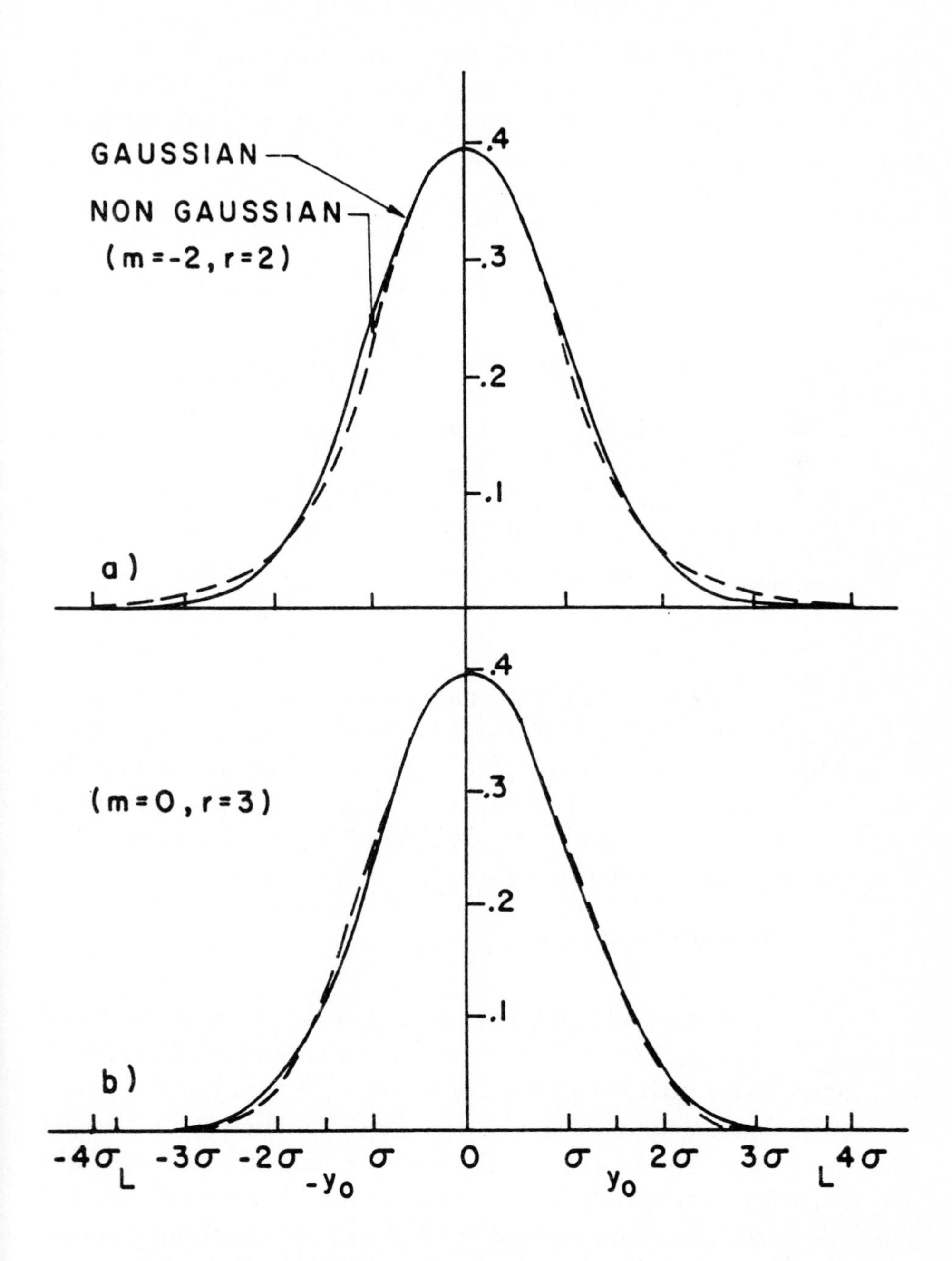

Figure 1.9: Normal distributions with uncertain tails.

non-Gaussian processes and a suitable number n of independent non-Gaussian variables. With such an equivalence postulated, Drenick constructs a non-Gaussian random process whose one-dimensional probability density $f(y)$ has the following properties:

1. Between two limits $(\pm y_0)$, $f_y(y)$ is Gaussian:

$$f_y(y) = \frac{a_1}{\sigma} \exp\left(-\frac{y^2}{2\sigma^2}\right) \quad , \quad (|y| < y_0) \tag{1.90}$$

2. Beyond these limits, $f_y(y)$ is of the exponential type

$$f_y(y) = \frac{a_2}{s} \left|\frac{y}{s}\right|^m \exp\left(-\frac{1}{2}\left|\frac{y}{s}\right|^r\right) \quad , \quad |y| > y_0, r > 1 \tag{1.91}$$

3. At $|y| = y_0$, $f_y(y)$ is continuous.

4. n is large.

5. L is large.

The first three assumptions are made in order to be able to postulate a random process which is Gaussian in a region in which observational data are available (namely for $y(t)$ values which are not very large) but which may depart from Gaussianity where such data are scarce and where such a departure would be difficult to ascertain statistically. The case of no departure is included. One merely sets:

$$r = 2 \quad , \quad m = 0 \quad , \quad a_1 = a_2 = \frac{1}{\sqrt{2\pi}} \quad , \quad s = \sigma \tag{1.92}$$

Suppose now also, as suggested above, that the exceedance probability $P\{\| y \| > L\}$ in eq.(1.87) for the random process y is the same as of n independent variables $y_1, y_2, \ldots y_n$, each with the normal density.

Under these assumptions, one can derive a formula for the exceedance probability $P\{\| y \| > L\}$ which is analogous to eq.(1.87). The derivation is laborious but straightforward. It is simplified if one can make a sixth assumption, namely $L \geq y_0 > s$, which is not unreasonable and which will be made here. One then finds

$$P\{\| y \| > L\} = 1 - \exp\left\{-\exp\left[-\frac{L}{s}(r \log 2n)^{\frac{r-1}{r}} + \eta'\right]\right\} \tag{1.93}$$

with

$$\eta' = \log a_2 + r \log 2n + \frac{m-r+1}{r}(\log r + \log\log 2n) \tag{1.94}$$

This is again valid asymptotically for large (L/s) and large n, but in the sense that terms of orders $O(L/s)^{-r}$ and $O(\log n)^{-1}$ are negligibly small.

Expression (1.93) for the exceedance probability is of roughly the same double exponential form as its counterpart, eq. (1.87). Since all parameters of the underlying density $p(y)$ enter into the second exponent, and some even exponentially so, the probability is very sensitive to even small changes in them.

The changes that are of interest here are those in the parameters a_1, a_2, s, m, and r, away from the values (1.92) which they take if the random process y is Gaussian. Their effect on the failure probability $P\{\| y \| > L\}$ can be evaluated by a conventional perturbation calculation. If P_f is used to denote the value of this probability when y is Gaussian, and δp the change induced by small departure, δs, δr, and δm from Gaussianity one finds

$$\frac{\delta P}{P_f} = [M_1 \frac{\delta s}{\sigma} + M_2 \delta m + M_3 \delta r] \log P_f \tag{1.95}$$

where

$$M_1 = \left(\frac{y_0}{\sigma}\right)^2 - 1 - \frac{L}{\sigma}\sqrt{2\log 2n} \tag{1.96}$$

$$M_2 = \log\frac{y_0}{\sigma} + \frac{1}{2}(\log 2 + \log\log 2n) \tag{1.97}$$

$$\begin{aligned} M_3 &= \frac{1}{4}\left(\frac{y_0}{\sigma}\right)^2\left(1 - 2\log\frac{y_0}{\sigma}\right) + \log 2n \\ &+ \frac{1}{4}\frac{L}{\sigma}\sqrt{2\log 2n}(1 - \log 2 - 4\log\log 2n) \\ &- \frac{1}{4}(1 + \log 2 + 4\log\log 2n) \end{aligned} \tag{1.98}$$

The expressions are valid if, as before, terms of orders $O(L/\sigma)^{-1}$ and $O(\log n)^{-1}$ are considered negligible relative to unity, and if the same is true of terms of order $O(y_0/\sigma)^{-2}$.

A mere inspection of Eq.(1.95) and (1.98) shows that even small changes in the one-dimensional density of the process y are prone to produce large changes δp in the failure probability. Numerical work confirms this (see Table 1.1). Suppose, for example, that a system had been designed on the assumption that y is Gaussian, and for a failure probability of $P_f = 0.05$. This would mean that L/σ would have been set at $L/\sigma = 1.64$. In order to simplify the formulas in eqs.(1.96)–(1.98), suppose further that $y_0 = L$ (i.e. that the departure from Gaussianity is most pronounced beyond the failure limit L), and that $n = 20$ (i.e. that the random process is equivalent to 20 independent random variables). In that case, one finds

$$\frac{\delta P}{P_f} = 15.3\frac{\delta s}{\sigma} - 22.4\delta r - 4.50\delta m. \tag{1.99}$$

This shows that merely a change in m alone from zero to unity, produces a change in the failure probability by a factor of 4.5

Such a change would be extremely difficult to detect statistically at the level of confidence which one would often wish to attach to an estimate of the failure probability. The usual statistical tests in particular which aim at the estimation of certain mean values of the density $f(y)$ of y, are known to yield no information regarding the behaviour for large values of y.

The evidence presented here therefore indicates that a reliable estimation of the failure probability will often be very difficult, basically of course, because it depends on the behaviour of the underlying random process for large values of its sample functions. That, however, is the region that is the least accessible to robust statistical tests.

Example. Propped Cantilever Imperfection Sensitivity. (Elishakoff (1983b)). Consider the cantilever of Fig. 1.10 — a rigid link of length a, pinned to a rigid foundation and supported by a linear extension spring of stiffness κ which is capable of resisting both tension and compression and which retains its horizontality as the system deflects. The initial imperfection is modelled by the deflection xa from the vertical position; the total displacement is denoted by ya.

Equilibrium dictates

$$\lambda y = (y - x)(1 - y^2)^{1/2}, \quad \lambda = \frac{P}{P_{cl}}, \quad P_{cl} = \kappa a. \tag{1.100}$$

	r				
m	1.0	1.5	2.0	2.5	3.0
− 2.0	0.213	0.333	0.352	0.268	0.150
− 1.5	0.389	0.394	0.171	0.072	0.022
− 1.0	0.432	0.215	0.081	0.023	0.005
− 0.5	0.372	0.140	0.040	0.009	0.0015
0.0	0.263	0.085	0.020[a]	0.004	0.0005
0.5	0.155	0.049	0.010	0.0016	0.0002
1.0	0.079	0.027	0.005	0.0008	0.0001
1.5	0.035	0.014	0.003	0.0004	0.00004
2.0	0.014	0.007	0.0014	0.0002	0.00002

[a] Gaussian case.

Table 1.1: Failure Probabilities $P\{\max_t |y_t| > L\}$.

For the nondimensional buckling load λ_s — the maximum load the structure can sustain — we put

$$\frac{d\lambda}{dy} = 0, \quad \lambda = \lambda_s. \tag{1.101}$$

Hence

$$y = x^{1/3}, \tag{1.102}$$

which upon substitution in eq.(1.100) yields an exact expression for the buckling load, due to Augusti (1964) and Thompson and Hunt (1973),

$$\lambda_s = (1 - x^{2/3})^{3/2}. \tag{1.103}$$

We assume next that the initial displacement is a uniform random variable X in the interval $(0, \xi)$ (Fig. 1.11). The buckling load is then itself a random variable Λ_s and

$$\Lambda_s = (1 - X^{2/3})^{3/2}. \tag{1.104}$$

We now calculate the reliability of the structure at the load level α, i.e. the probability of it not buckling below this load α,

$$R(\alpha) = \text{Prob}(\Lambda_s > \alpha) = \text{Prob}\left[\left(1 - X^{2/3}\right)^{3/2} > \alpha\right] \tag{1.105}$$

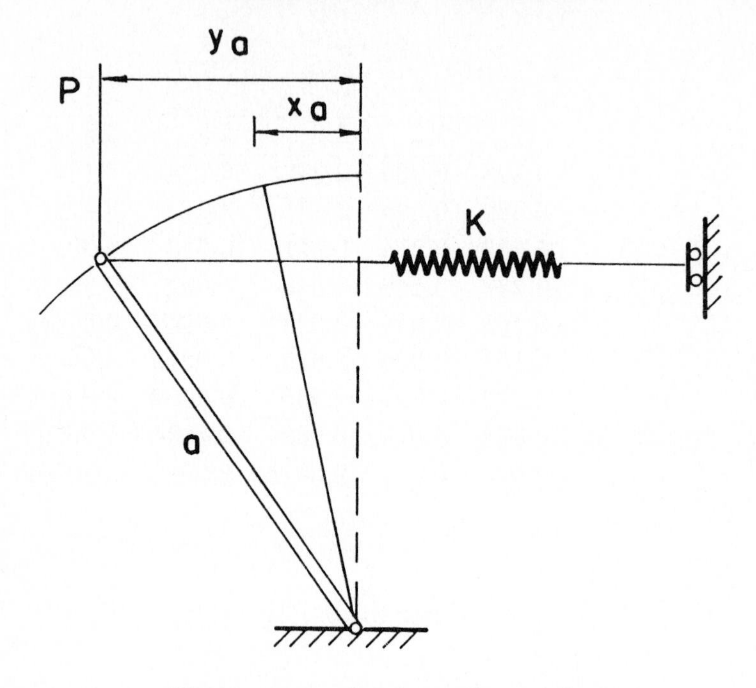

Figure 1.10: A propped cantelever.

$$= \text{Prob}\left[X < \left(1 - \alpha^{2/3}\right)^{3/2}\right] \tag{1.106}$$

Hence

$$R(\alpha) = \begin{cases} 0 & \text{for} \quad \alpha > 1 \\ \frac{1}{\xi}\left(1 - \alpha^{2/3}\right)^{3/2} & \text{for} \quad \alpha^* < \alpha < 1 \\ 1 & \text{for} \quad \alpha < \alpha^* \end{cases} \tag{1.107}$$

where

$$\alpha^* = \left(1 - \xi^{2/3}\right)^{3/2}. \tag{1.108}$$

Say that the acceptability of the structure consists of the requirement that the reliability be not less than some required level $r < 1$. The allowable load is then a minimum root of

$$R(\alpha) = r \tag{1.109}$$

and is given by

$$\alpha_{\text{allow}} = \left[1 - (\xi r)^{2/3}\right]^{3/2}. \tag{1.110}$$

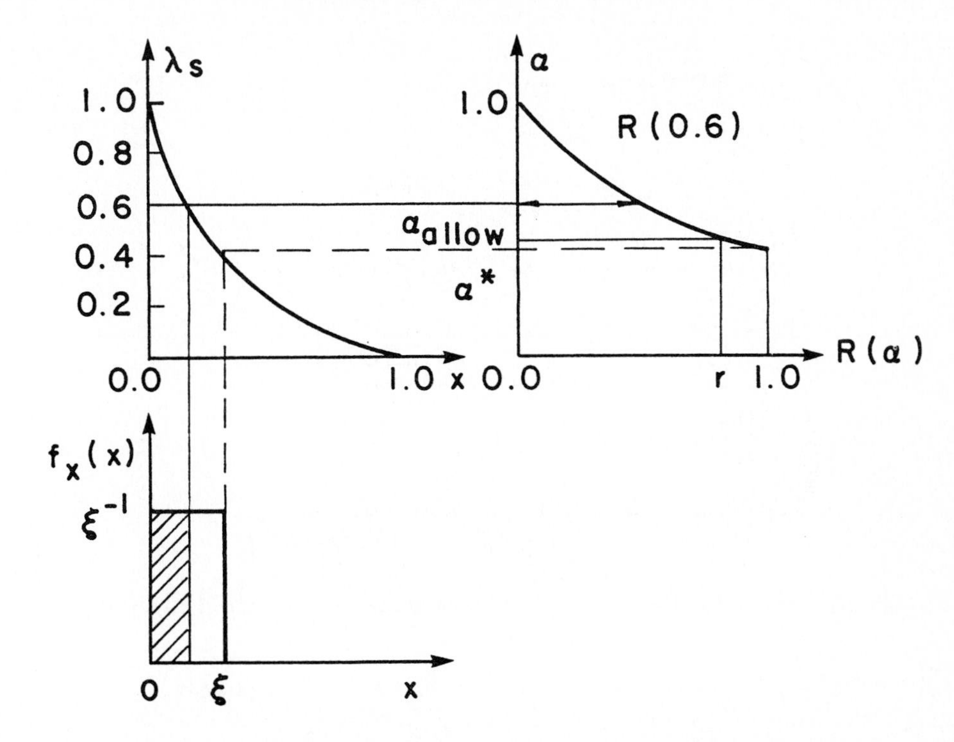

Figure 1.11: Reliability for uniform distribution of X.

Curve 1 in Fig. 1.12 shows α_{allow} versus r for $\xi = 0.3$. As is seen in the high-reliability range, α_{allow} is a flat function of r.

An exact buckling load–imperfection amplitude relationship λ_s–versus–x is rarely available, and usually approximate analysis is used. We now consider the effect of an approximate relationship on the reliability estimate. Two-term approximation in Eq.(1.104) yields

$$\lambda_s = 1 - \frac{3x^{2/3}}{2} \tag{1.111}$$

with the corresponding reliability estimate

$$R^{(e)}(\alpha) = \begin{cases} 0 & \text{for} \quad \alpha > 1 \\ \frac{1}{\xi}\left[\frac{2(1-\alpha)}{3}\right]^{3/2} & \text{for} \quad \alpha^{**} < \alpha < 1 \\ 1 & \text{for} \quad \alpha < \alpha^{**} \end{cases} \tag{1.112}$$

where

$$\alpha^{**} = 1 - \frac{3}{2}\xi^{2/3}. \tag{1.113}$$

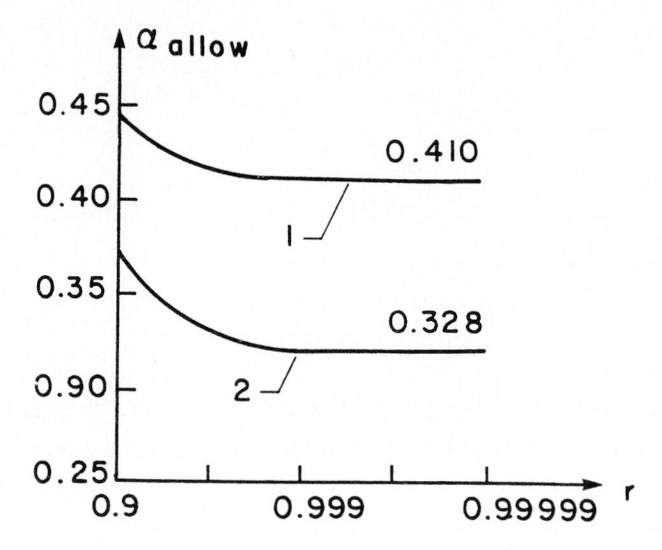

Figure 1.12: Allowed values of the load, α, versus the reliability, r.

The associated approximation of the allowable load equals

$$\tilde{\alpha}^*_{\text{allow}} = 1 - \frac{3}{2}(\xi r)^{2/3} \tag{1.114}$$

and is shown on curve 2 in Fig. 1.12. As is seen, the difference between the exact and approximate allowable loads may reach 20% for the case under consideration.

As regards the effect of an error δ in the probability density of the initial imperfection amplitude we assume that

$$\begin{aligned} p_X(x) &= \frac{1}{\xi+\delta} \quad \text{for } x < \xi + \delta \\ &= 0 \quad \text{otherwise} \end{aligned} \tag{1.115}$$

The resulting allowable load is then

$$\alpha_{\text{allow},\delta} = \{1 - [(\xi+\delta)r]^{2/3}\}^{3/2}. \tag{1.116}$$

By contrast, the value of the *exact* reliability at this load is

$$\begin{aligned} R(\alpha_{\text{allow},\delta}) &= \left(1 + \tfrac{\delta}{\xi}\right) r \quad \text{for } \delta < 0 \\ &= 1 \qquad\qquad \text{for } \delta > 0 \end{aligned} \tag{1.117}$$

On the basis of inexact experimental information on $f_X(x)$ (e.g. for $\delta/\xi = -0.01$), the impression could be that the design refers to very high reliability, say $r = 0.999\,999\,9$, but in fact it cannot exceed 0.99. On the other hand, the inaccuracy may occasionally be "useful": at $\delta > 0$, the exact reliability is unity, whereas the approximate allowable load corresponds to $r < 1$. The lesson learned from this example is that the required reliability must be compatible with the accuracy of the input data.

Example. Random Loading. Consider now how randomness of the applied load influences the reliability of the structure. There is a claim in the literature that such randomness (besides that of the initial imperfections) should reduce reliability. To elucidate this question, we will consider the case where the applied load is a random variable Λ, uniformly distributed over the interval (λ_1, λ_2), that is,

$$p_\Lambda(\lambda) = \begin{cases} \dfrac{1}{\lambda_2 - \lambda_1} & \text{for} \quad \lambda_1 < \lambda < \lambda_2 \\ 0 & \text{otherwise} \end{cases} \tag{1.118}$$

Under the new circumstances the reliability will be defined as

$$R = \text{Prob}(\Lambda_s > \Lambda) \tag{1.119}$$

With the quite reasonable assumption of independence of Λ_s and Λ (due to independence of the manufacturing process and applied loads) we find (by analogy with Eq. (7.4) of Elishakoff, 1983)

$$R = \int_0^\infty [1 - F_{\Lambda_s}(z)] p_\Lambda(z)\, dz \tag{1.120}$$

where $F_{\Lambda_s}(z)$ is the distribution function of the buckling loads. Note that $1 - F_{\Lambda_s}(z)$ coincides formally with eq.(1.107) with α replaced by z, and integration in the above formula has to be carried out accordingly. Consider, for example, the case which corresponds to required reliability r where the applied force is deterministic. The allowable load is as per eq.(1.110). Assume now that the applied load is a uniformly distributed random variable, with $\lambda_1 = 0$ and $\lambda_2 = \alpha_{\text{allow}}$. Then

$$R = \frac{1}{\alpha_{\text{allow}}} \left[\int_0^{\alpha^*} dz + \frac{1}{\xi} \int_{\alpha^*}^{\alpha_{\text{allow}}} \left(1 - z^{2/3}\right)^{3/2} dz \right] \tag{1.121}$$

where α^* is given by eq.(1.108). Calculation yields

$$
\begin{aligned}
R &= \frac{\alpha^*}{\alpha_{\text{allow}}} - \frac{3}{\alpha_{\text{allow}}\xi}\left[\frac{3\beta_2}{48} + \frac{\cos\beta_2 \sin^5\beta_2}{6} - \frac{\sin 2\beta_2}{24} + \frac{\sin 4\beta_2}{192}\right. \\
&\quad \left. - \left(\frac{3\beta_1}{48} + \frac{\cos\beta_1 \sin^5\beta_1}{6} - \frac{\sin 2\beta_1}{24} + \frac{\sin 4\beta_1}{192}\right)\right] \qquad (1.122)
\end{aligned}
$$

where

$$\beta_1 = \arccos(\alpha^*)^{1/3} \quad , \quad \beta_2 = \arccos(\alpha_{\text{allow}})^{1/3} \qquad (1.123)$$

For example, for $r = 0.999$ and $\xi = 0.3$, we have $\alpha_{\text{allow}} = 0.4103$, $\alpha^* = 0.41$ and $R = 0.999\,999\,5$. That is, the reliability of the structure with random applied load exceeds that of the structure under the deterministic load considered. However, if $\lambda_1 = 0.4103$ and $\lambda_2 = 0.4103 \times 1.01 = 0.4144$, the reliability will be

$$R = \frac{1}{(\lambda_2 - \lambda_1)\xi} \int_{\lambda_1}^{\lambda_2} (1 - z^{2/3})^{3/2}\, dz = 0.992\,843\,9 \qquad (1.124)$$

which is less than r associated with the deterministic load. This implies that randomness of the additional parameter may increase or decrease the estimate of the actual failure probability.

1.3 Some Quotations on the Limitations of Probabilistic Methods

> The mathematics of chance encompasses all of our life: inorganic, organic and social. It is a science and it is an art, since life itself is "the art of drawing sufficient conclusions from insufficient premises" [Samuel Butler][5]

The validity of probabilistic analysis has long been a subject of throught. Though the probabilistic approach to various problems of

[5]Kordemski, 1981.

theoretical and applied mechanics achieved a high degree of sophistication, it appears that it has some limitations. We supplement our previous discussion with quotations from a range of distinguished workers who have contributed to both sides of the dispute over the role of probabilistic models. Our view of modelling, presented at the beginning of the chapter, suggests that models of uncertainty of both sorts — probabilistic and non-probabilistic — are relevant in appropriate circumstances. The aim of the debate presented in the remaining pages of this chapter is not to resolve once and for all the validity of one approach or another (though some of the authors to be quoted would disagree). Rather, the goal is to try to elucidate the conditions under which the probabilistic approach is valid.

It is only proper to open with Feodosiev's powerful and spirited discussion of the possible limitations of probabilistic methods, quoted from his informal lecture–conversations (Feodosiev, 1969):

> The parameters of external action are not strictly permanent. There is always some probability of their deviating from the nominal value. The same is the case with the internal ones: there is scatter of the mechanical characteristics of the material — nonuniformity of its properties due to the production technology, to deviations in geometry, etc.
>
> With the above deviations treated as random and occurring within some range, the probability of failure of a structure can be predicted with the aid of the law of large numbers. When this probability turns out to be practically zero or in any case very low, the structure can be regarded as reliable.
>
> Before discussing the practicality of such an approach, we ought to consider the legitimacy of the probabilistic conception itself, which treats failure as a random event. This is by no means obvious — not at all times, not to everyone.
>
> Imagine a fairly typical dialogue between an adherent of the probabilistic conception of strength, and some strict opponent. Without going into the details of the proposed analysis, the opponent asks point-blank: 'How can we, when building a dwelling house, envisage — nay, calculate and plan — some probability of its collapse? How can we, with such an approach, judge the output of the designer and the builder? And most important of

all, how can we take seriously the degree of their responsibility for their output?'

This is countered by their argument that the probability of any building collapsing exists regardless of whether we "like" it or not. It is an objective characteristic. It is very low but still above zero, and it manifests itself in the practical fact that even in the housing sector, collapses occur once in 10-20-30 years, both during the construction stage and after completion. Cases of partial failure are still more numerous.

'True', says the opponent. 'But you will agree that your examples refer not to the probability of scatter within the limits of the prescribed technological requirements, but to that of their being violated: that of faulty technology, of acceptance of substandard materials — finally, of actual bungling. Naturally, statistical recording of such events would be as useful as, say, that of street accidents. This, however, is the business not of the engineer but rather of the work-organization sociologist.'

'Quite so', replies the probabilist. 'My example of collapse during construction is indeed irrelevant. But let the structure be realized under total observance of the requirements. Even then the probability of collapse differs from zero. This is especially well reflected in the context of external loads, which are known only as a statistical estimate. Take again the dwelling house as an example. In considering its strength with reference to seismic action, we rely on the statistics of earth tremors in the region in question. Near rivers, we rely on the long-term statistics of flood levels. The same is the case with wind loads.'

With this reasoning, the opponent is in agreement. But something else bothers him.

'Assume', he says, 'that the external loads are known to precision. Let us consider the technological requirements regarding production. If they are correct and complied with in full, I would say that the probability of collapse is zero. Stagewise production control is actually practised with a view to early exclusion of such deviations from the nominal as might lead to a failure situation. If the probability of collapse is still preserved, what this means is that the prescribed control is faulty.'

This argument is readily refuted. The probabilistic approach is in fact advocated in order, inter alia, to evaluate the control procedure; to give it scientific grounding; to tighten it where necessary, to relax it where possible.

'Tell me', says the opponent, 'such an analysis necessitates a large volume of statistical production data. I have a reason for asking. So far, I have not seen a single calculation of the simplest structure based on statistical processing of production factors. Worse, I feel that there exists currently a definite "passion" for the mathematical side of the problem, a turning-away from the search for the initial parameters. The necessary information is simply not available, with no clue as to how to look for it. The answer here is based on proper faith in technological and scientific progress.'

'The difficulties in obtaining initial data are obvious. We can rest assured, however, that they are surmountable; with improved — and proliferating — control and measurement tools, this aspect would be solved. This, at least, has been the case with all applications of probabilistic procedures in various fields of technology.'

Before we take leave of our disputants, let us hear the opponent's concluding reply, which shows that his doubts have not been allayed: 'In technology, failure is commonly regarded not as a random event but as an extraordinary occurrence. To avoid it, all stages of production are accompanied by control, testing and modelling. All these activities are subject to continual modification and improvement. It is simply impossible to keep up with them from a probabilistic position. Who is to do it? What would be the cost of compiling the necessary data from which you might perhaps be able to compute the theoretical probability of failure? But that is not the main point. I repeat, in engineering practice it is not so much the probability of scatter within the requirement limits that counts, as that of deviation from the latter. This is the overall determining factor. An engineer responsible for a structure is worried first and foremost — albeit regrettably — not about the probability of scatter you are interested in, but about practical measures ensuring quality. And let us not hold this against him. His problem is more complicated than yours. On occasion he has to play the psychologist. To which master welder should he assign an important joint, at what hour of the day and even on which day of the week or month? ... You might try to fit this reality into your analytical scheme .... '

So let us break off this discussion. From the pro and con arguments advanced, much becomes obvious.

Where external parameters are involved (loads, temperatures, etc.) the statistical approach is not merely legitimate; it is realizable and in some cases the only one possible. Such are the seismic loads referred to earlier, and also the wind loads and the loads which serve — or should serve — as a basis in transportation design.

As regards the probabilistic approach to internal parameters, the matter is more complicated. There is a range of problems which cannot be tackled properly without recourse to the specifics of human activity in the course of the lengthy sequence of production stages — from smelting of the metal to exploitation of the finished structure. Inclusion (or exclusion) of these specifics could perhaps be successfully achieved in particular cases, but to accept total predominance of the probabilistic conception seems unreasonable — even for the distant future.

Having renounced the treatment of failure as a probabilistic event, strength analysis of structures is considerably simplified and acquires realistic grounding.

... As in the matter of strength, the attempt to base stability analysis on the probabilistic approach is not new. The problem can be posed under the assumption of negligibility of the initial imperfections, with only the perturbations associated with the loading process regarded as probabilistic.

The potentialities of such an approach are fairly limited. There is an analogy. Water discharged in a watershed would flow in one direction or another according to the initial conditions of the watershed in question, not according to subsequent perturbations; thus it is the initial deviations from the nominal that decide the outcome.

Attempts at statistical processing of initial imperfections have little chance of success at present. This was adequately discussed in the context of the probabilistic approach to strength problems. The grounds for such a prognosis are essentially the same.

This should be borne in mind in view of the fact that the

significance of the probabilistic approach is often exaggerated, and there is very real danger of the effort invested in urgent creation of mathematical tools for predicting loss of stability as a random event — proving to be wasted, when the indispensable distribution functions of the imperfections remain unknown even in those few cases where they can be categorized as random parameters. The practicing engineer is certain, in doubtful cases, to prefer stricter quality control, or even changes in design, to costly study of "transient" distribution functions.

As we see, the main reservation against the probabilistic approach is the need for assumptions (often not supported by experimental evidence) as to the probabilistic nature of the involved uncertain quantities.

Assumption of a Gaussian (normal) distribution is common practice. For example, as Hahn and Shapiro (1967) mention:

> The normal is the most widely used of all distributions. For a long time its importance was exaggerated by the misconception that it was the underlying distribution of nature, and that according to the "Theory of Errors" it was supposed to govern all measurements. With the advent of statistical tests about the year 1900, this assumption was shown not to be universally valid.
>
> Instead, the theoretical justification for the role of the normal distribution is the *central limit theorem*, one of the most important results of mathematical statistics. This theorem states that the distribution of the *mean* of n independent observations from *any distribution*, or even from up to n different distributions, with finite mean and variance approaches a normal distribution as the number of observations in the sample becomes large — that is, as n approaches infinity. The result holds, irrespective of the distribution of each of the n elements making up the average.
>
> Although the central limit theorem is concerned with large samples, the sample mean tends to be normally distributed even for relatively small n as long as no single element or small group of elements has a dominating variance and the element distributions do not deviate extremely from a normal distribution.
>
> When a random variable represents the total effect of a large number of independent "small" causes, the central limit theo-

> rem thus leads us to expect the distribution of that variable to be normal. Furthermore, *empirical evidence* has indicated that the normal distribution provides a good representation for many physical variables. Examples include measurements on living organisms, molecular velocities in a gas, scores on an intelligence test, average temperatures in a given locality, and random electrical noise. Instrumentation error is also often normally distributed either around the true value or around some average bias. The normal distribution has the further advantage for many problems that it is tractable mathematically. Consequently, many of the techniques of statistical inference, such as the method known as the "analysis of variance", have been derived under the assumption that the data come from a normal distribution.
>
> Because of the prominence, and perhaps the name, of the normal distribution, it is sometimes assumed that a random variable is normally distributed unless proven otherwise. Therefore it should be clearly recognized that many random variables *cannot* be reasonably regarded as the sum of many small effects, and consequently there is no theoretical reason for expecting a normal distribution. This could be the case when one nonnormal effect is predominant. ...
>
> ...Finally, it should be noted that for some random variables a normal distribution provides a reasonable approximation in the center, but *is inadequate at one or both tails of distribution.*[6]

However, the probability of failure of the structure is mainly determined by these tails of the distribution. Etkin (1980) indicates, that

> Although there is much evidence that turbulence is not in fact a Gaussian process, with small and large values both occurring more frequently than in a normal distribution, the assumption that individual patches are Gaussian is widely used because of the great analytical advantage it offers.

This statement is very similar to that of Ottestad (1970):

> An important premise for the mathematical deduction was that the observed random variable is normally distributed. There

[6]The italics are due to Hahn and Shapiro.

> are several grounds for doubting the realism of this premise. It is hardly possible that any random variable exists which is exactly so distributed. Certainly, a large number of actual random variables are found, the distributions of which closely resemble the normal form, but there are also actual distributions that deviate considerably from this model.

Blekhman, Myshkis and Panovko (1983) emphasize that

> Significantly, the weakness of numerous works on stochastic models — sometimes ruling out any application — lies in the choice of statistical hypotheses, especially of assumptions regarding the probabilistic features of the given accidental quantities and functions. These features are often regarded as fully known (e.g. assumption of a normal distribution with known parameters), or as capable of determination.
>
> In real situations, however, it mostly turns out that the needed information is lacking; worse, in many cases the situation turns out to be not stochastic but indeterminate, so that the very applicability of statistical hypotheses is in doubt.

The following quotation belongs to Leontief (1971) in his Presidential address delivered at the 83rd meeting of the American Economic Association. While originally directed towards economic issues it has some relevance in engineering:

> The validity of these statistical tools depends itself on the acceptance of certain convenient assumptions pertaining to stochastic properties of the phenomena which the particular models are intended to explain; assumptions that can be seldom verified.
>
> Theorists continue to turn out model after model and mathematical statisticians to devise complicated procedures one after another. Most of these are relegated to the stockpile without any practical application or after only a perfunctory demonstration exercise.

Smith (1986) in his textbook also refers to this problem:

> The mathematics involved in the application of full probability theory are esoteric and, not unnaturally, most civil engineers did not welcome any moves which might lead to the adoption of such methods in their work.

Simiu and Scanlan (1986) in their monograph on the wind effects on structures, stress that

> Owing to physical and probabilistic modelling difficulties and to the absence of sufficient statistical data it is in general not possible to provide confident probabilistic descriptions of the loads, particularly within the range corresponding to ultimate limit states. Comprehensive probabilistic descriptions of the relevant physical properties of the structure are also seldom available.

Kogan (1976) in his book on crane design notes that

> A detailed statistical analysis of the data obtained by 113 weather stations in France on the maximum wind speed over a period of 23 years, carried out by Duhene-Marullaz (1972), showed that the distribution laws of Gauss, Gumbel, and Frechet are equally applicable, although they all yield large errors at high wind speeds.

Blekhman, Myshkis and Panovko (1983) again emphasize that

> In the theory of probability the probabilistic characteristics are actually primary — like lengths and angles in geometry — so that it would seem logical, in resorting to a stochastic approach, to regard them as given. The aim of applied mathematics, however, is not merely to determine some quantities from others (or, in particular, logically secondary quantities from logically primary ones) — but to find, from quantities which may realistically be regarded as given (i.e. which are capable of direct measurement or calculation), those incapable of direct determination. Thus, in constructing a mathematical model, the question of obtainability of initial data, and of the effort involved, is of paramount — possibly decisive — importance.

Bolotin (1968) provides a powerful reply to the above reservations and strongly defends the probabilistic approach:

> Let us refer briefly to the reservations advanced in both the past and present by opponents of the probabilistic and statistical approaches. These considerations reduce essentially to two

basic arguments. The first is doubt as to the feasibility of collecting experimental data in sufficient amounts for probabilistic processing. This doubt, while perhaps valid in the past, need no longer be entertained. Advances in automatics and metrology, permitting automatic recording and actual planning of the experiment itself; widespread recourse to computers, permitting rapid statistical processing of large volumes of data — all these have obviated not only the basic but also the practical difficulties ....

The second argument advanced against the probabilistic approach is of a more abstract nature. There are those who feel that probabilistic conclusions are only relevant to mass events and to systems reproduced as large populations and exploited under uniform conditions. In other words, it is claimed that probabilistic reasoning is valid only where probabilities lend themselves to statistical interpretation and the law of large numbers applies. However, probability is an objective measure of the "feasibility" of an event; a measure which retains its sense irrespective of the event being multiply reproducible or otherwise, and which is used by us (albeit semi-intuitively) in everyday practice in gauging the chances of some situation or other occurring. This approach acquired scientific grounding in the theory of operations — an applied discipline whose aim is sound planning of actions with a view to a probabilistically optimal effect. The probability of a projected system performing properly over a specified service life remains an objective measure of its reliability even if the system is realized as a single specimen. This probability may be used, for example, for comparison against some normative counterpart adopted on the basis of current design practices, as well as for mutual comparison of a different version of the system in question.

As a rule, the forces acting on a structure admit multiple reproduction, or develop their probabilistic features in time. Structural materials are mass-produced, and their mechanical properties in different batches lend themselves to exhaustive study. Composite structural elements are usually likewise of mass type and in any case may be produced in sufficient quantities for statistical conclusions. Thus the behavior even of a single-specimen system is determined, in the final analysis, by

> accidental mass-type factors, each of which is subject to statistical interpretation of probabilities and to the law of large numbers. Prediction of this behavior on the basis of such statistical data is, in effect, the aim of statistical mechanics and of the theory of reliability.
>
> As an "antithesis" to the probabilistic approach, methods are occasionally put forward based on the concepts of "infrequently-occurring", "maximum" and "minimum" loads and strengths. In reality these are, however, no more than substitute probabilistic procedures dispensing with the theory of probability — and in spite of their apparent simplicity and obviousness, entailing irremediable logical contradictions. They are inapplicable in practice without arbitrary decisions — which render them implausible and inadequate to a large degree.
>
> Thus, recourse to the methodology of statistical mechanics and reliability theory calls for drastic increase in the volume of data on external forces (and on the environment in general), as well as on materials. This increased volume of essential information is the natural price to be paid for more exact prediction of the behavior of a structure, and for more authentic conclusions regarding its reliability and service life.

The natural question arises: Who is right? Those who are criticizing probabilistic methods or those who go out to defend them? Paradoxically, it appears that both answers are correct.

When sufficient information is available to enable a reliable deterministic analysis, ample techniques are accessible. Alternatively, where the necessary information to formulate a stochastic model is at hand, it is fully justifiable to apply probabilistic methods. Indeed, the utility and beauty of probabilistic methods in engineering have been clearly demonstrated in the last twenty years. The situation is different if we do not possess sufficient probabilistic information. Often only very limited knowledge is available and appropriate mathematical tools are needed.

For example, if one has a cylindrical shell which is part of a critical structure (for example a liquid booster in the space shuttle) then full initial imperfection measurements of its shape could be performed and this data could be used as input to powerful numerical codes for calculating the buckling load. (Such measurements of a 10-foot diameter shell were performed by Arbocz (1982) at the NASA Langley Research Center). If, on the other hand, a large collection of shells is available, one could study the initial imperfection profiles statistically. (An analysis such as this, which directly incorporates experimental information in the probabilistic design rather than making assumptions on the nature of this information, is given by Elishakoff and Arbocz (1985)).

There is, however, an intermediate situation where only partial knowledge of the uncertain parameters and functions is available. In these circumstances one should apply the new analysis: convex modelling. The fundamentals of this approach are the subject of the following chapters.

Chapter 2

Mathematics of Convexity

> Those who boast of creating mathematics untainted by the physical world and maintain under pressure that others will some day find justification for their currently pointless endeavors may be left to work in their own mental grooves. But they contradict the entire course of history. Their confidence that a mathematics freed from bondage to science will produce richer, more varied, and more fruitful themes that will be applicable to far more than the older mathematics is not backed by anything but words.[1]

2.1 Convexity and Uncertainty

The elementary properties of convex sets provide the basic tools for much of our discussion of uncertainty in this book. The aim of the present chapter is to briefly review the mathematical properties of convexity. Before embarking on this task, however, it is appropriate to present a preliminary discussion of the importance and usefulness of convex sets in representing uncertainty.

The probabilistic approach to the modelling of uncertainty begins by defining a space of events and a probability measure on that space.

---

[1] Kline, 1980, pp 300 – 301.

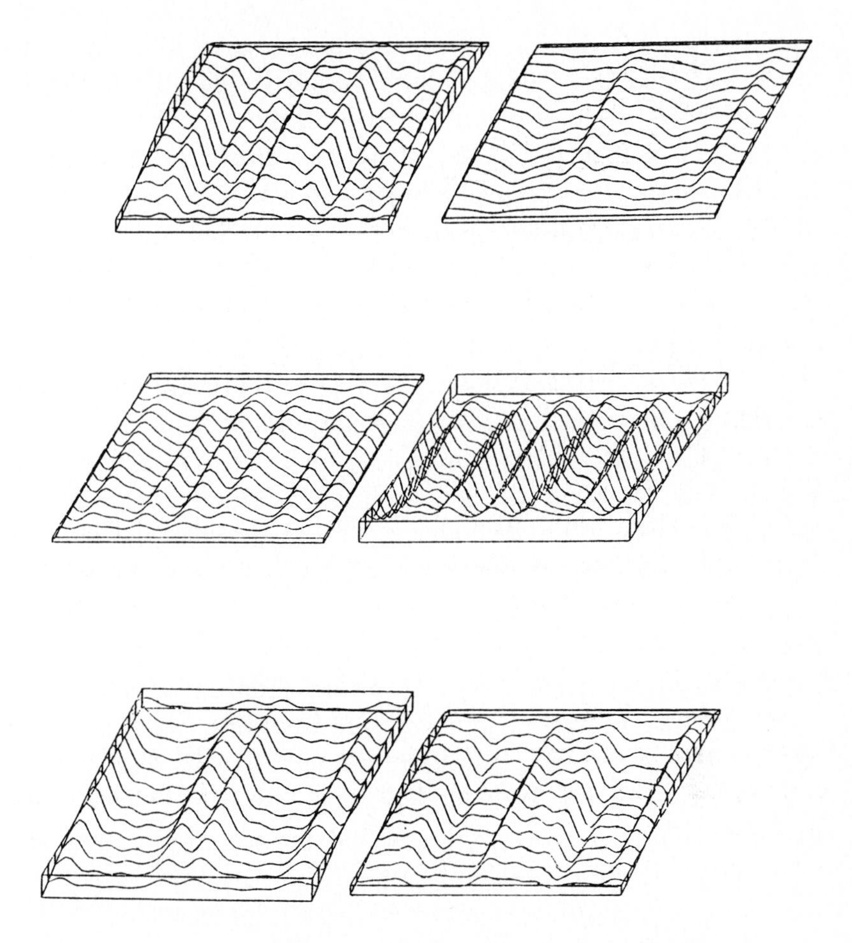

Figure 2.1: General geometrical imperfections of shells.

The space is all-inclusive; everything that could occur, and also possibly events that cannot occur, are included. The probability measure contains all information concerning the relative frequency of different events.

The set-theoretic approach to the modelling of uncertainty is different. A space of conceivable events is defined, as in the probabilistic approach. However, no probability measure is defined. Rather, sets of allowed events are specified, and the structure of these sets is chosen to reflect available information on what events can and cannot occur.

As a simple example, consider the geometrical imperfections of a cylindrical shell. Fig. 2.1 shows planar views of simulated shell imperfections (Elishakoff, 1982). These shapes range from fairly smooth and regular ridges to highly convoluted and uneven distortions. The space of possible shapes is the set of all two-dimensional surfaces with no self-intersections. The probabilistic approach to modelling the uncertainty in the shell shape is to define a function on the set of possible shapes, which gives the probability density for each particular shape. By assuming a Gaussian model for the shell imperfections, the probability density can be related to a correlation function of the surface imperfections. Alternatively, the probabilistic approach may define a probability density function whose argument is a vector containing a finite number of spatial dimensions or other parameters of the shell shape. The set theoretic approach is to define subsets of the space of possible shapes. The structure of the sets of allowed shapes will represent information on the range of shapes which occur. A family of sets can be constructed to represent different degrees of variation of shape, in connection with various constraints such as age, manufacturing conditions, material, etc. Likewise, different classes of sets can be formulated to represent different types of information on the allowed shapes. For example, a simplistic model supposes that the geometrical imperfections of the shell surface vary arbitrarily between lower and upper bounds. A more sophisticated model may include information on variation of both the displacement and the curvature of the surface.

These considerations can immediately be extended to the formulation of set-theoretic models of uncertainty in a wide range of situations. The remainder of this book is devoted to a number of examples in the field of applied mechanics. The selection of a set-theoretic representation of uncertainty is motivated, in each case, by the particular details of the application in question. However, it is remarkable, and of considerable practical significance, that sets whose elements represent spatial or temporal uncertainty are often found to be convex. Before embarking on a discussion of set-models of uncertainty for specific engineering applications we wish to identify some conditions in which it is plausible to assume such sets of functions are convex.

The central limit theorem will motivate our discussion. Let $g_1, \ldots,$ g_n be independent, identically distributed random variables with zero mean and finite variance. As $n \to \infty$ the distribution of the sum

$f = \frac{1}{\sqrt{n}} \sum g_i$ tends to a normal distribution, regardless of how the g_i are distributed. The physical analog of this theorem suggests that if a certain measureable *macroscopic* quantity f — e.g. a voltage or a temperature — is the superposition of numerous random, independent and identically distributed *microscopic* variables g_i, then we should expect the macroscopic quantity f to display a Gaussian distribution, regardless of how the g_i are distributed. Indeed, this expectation is fulfilled in many circumstances.

Now let us consider a set-theoretic approach to modelling the uncertainty of a time-dependent macroscopic vector function f. Let Γ be a set of vector-valued functions. For a positive integer n, consider the set of functions:

$$F_n = \left\{ f : f(t) = \frac{1}{n} \sum_{i=1}^{n} g_i(t) \text{ for } g_i \in \Gamma, i = 1, \ldots, n \right\} \tag{2.1}$$

F_n is the set of all n-fold averages of vector functions in Γ. It is well known (Aumann, 1965; Artstein, 1974; Artstein and Hansen, 1985) that, as $n \to \infty$, the sequence of sets F_n converges to the convex hull of Γ. This result invites the following physical interpretation. If a *macroscopic* time-dependent vector $f(t)$ is formed as the superposition of numerous *microscopic* time-varying events $g_i(t)$ chosen from a set Γ, then the set of all such functions $f(t)$ will tend to be convex, regardless of the structure of the set Γ.

These considerations suggest that vector functions representing complex and uncertain processes will tend to cluster into convex sets of functions. In the course of our discussion in the following chapters we will encounter numerous practical examples. We are now ready to begin a brief survey of pertinent concepts relating to convexity.

2.2 What is Convexity?

2.2.1 Geometric Convexity in the Euclidean Plane

A region is convex if the line segment joining any two points in the region is entirely in the region. See Fig. 2.2. The regions S and T are convex, while U and V are not. Circles and triangles delimit convex

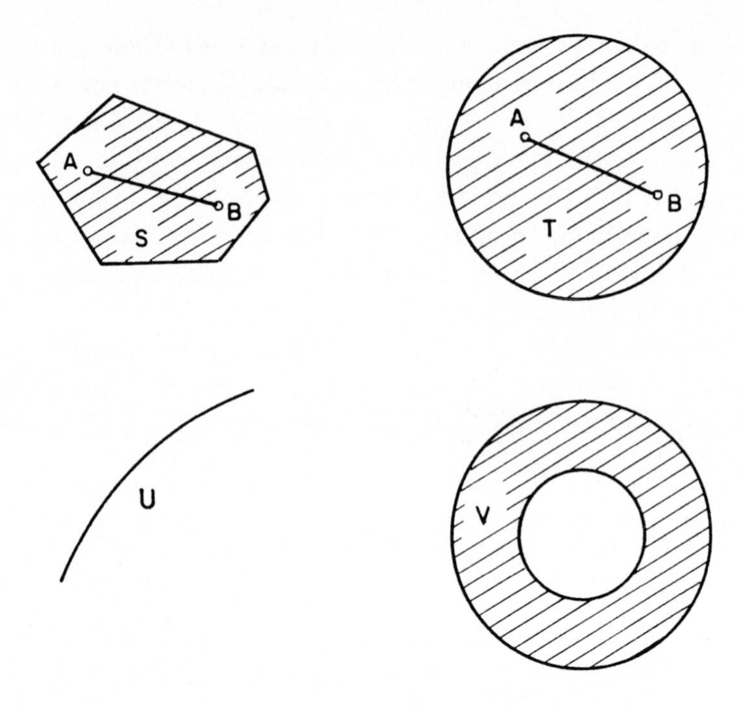

Figure 2.2: Convex and non-convex regions.

regions whereas quadrilaterals may or may not, depending on whether their diagonals intersect within the region.

The *intersection* of two sets is the set which contains all the points belonging to both sets. The *union* of two sets is defined as the set containing all the points belonging to either set. The intersection of convex sets, if it is not empty, is convex. This is true because the convexity of a region is a property of its boundary. However, it is evident that the *union* of convex regions is not necessarily convex. See Fig. 2.3, where the intersection of the triangular regions A and B is convex, but their union is not.

The concept of convexity has engaged the minds of thinkers since the emergence of western mathematics (Fenchel, 1983). Archimedes, in his treatise "On the Sphere and Cylinder" defined a convex arc as a plane curve which lies on one side of the line joining its end-points and all chords of which lie on the same side of the curve. In his treatise "On the Equilibrium of Plane Figures" Archimedes studies the centroids of

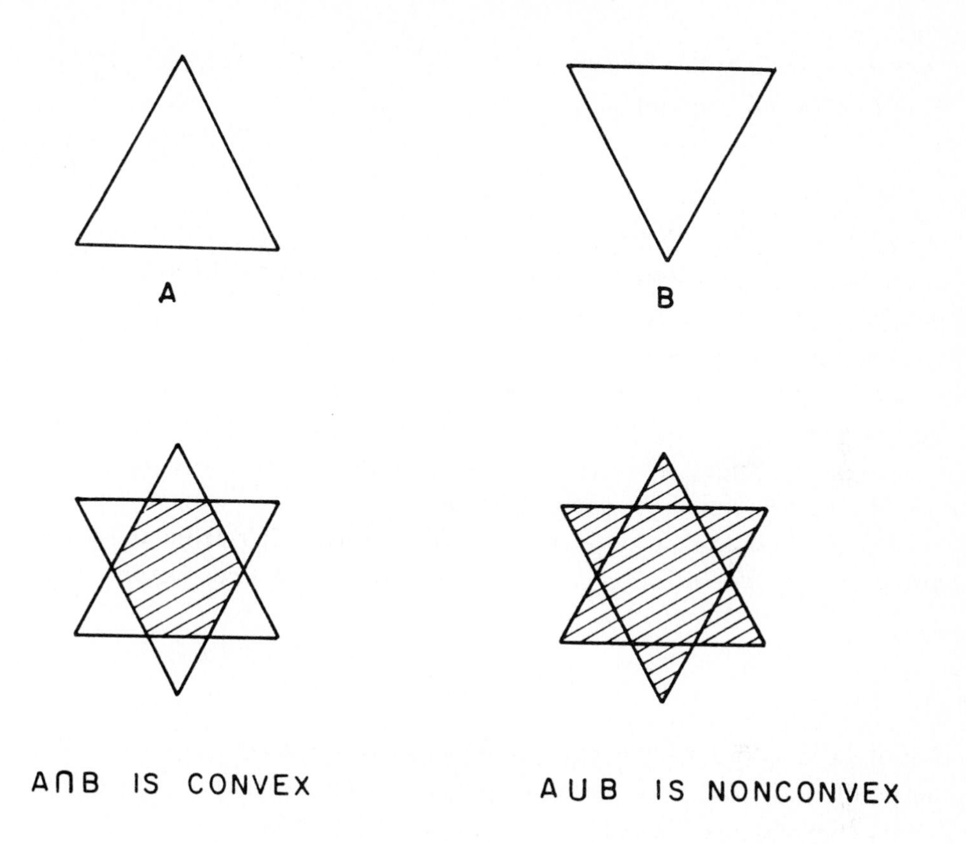

Figure 2.3: Unions and intersections of convex regions.

plane figures and concludes that a convex figure contains its centroid.

2.2.2 Algebraic Convexity in Euclidean Space

In order to exploit these and other properties of convex sets, we need an algebraic formulation of convexity. Nevertheless, the geometric presentation will sometimes be a useful intuitive guide.

Let S be a set of points in N-dimensional Euclidean space, E^N. Then S is convex if all averages of points in S also belong to S. That is, S is convex if, for any points $p \in S$ and $q \in S$, and any number $0 \leq \alpha \leq 1$

$$\alpha p + (1 - \alpha) q \in S \tag{2.2}$$

To see that this definition is equivalent to the geometrical definition of convexity, let p and q be two points in E^N. The line segment joining p and q is the set of points

$$r = \alpha p + (1 - \alpha)q \tag{2.3}$$

for all $0 \leq \alpha \leq 1$. Thus a set is convex by the algebraic definition if it contains all line segments whose end-points are in the set, which is precisely the geometric definition.

An expression such as $\alpha p + (1 - \alpha)q$, for $0 \leq \alpha \leq 1$, is called the *convex combination* of the points p and q. Likewise, if $p^1, \ldots, p^m$ are points and $\alpha_1, \ldots, \alpha_m$ are non-negative numbers which sum to unity, then $\alpha_1 p^1 + \ldots + \alpha_m p^m$ is the convex combination of these points, and $\alpha_1, \ldots, \alpha_m$ are called *convex coefficients.*

Let us show that a hyperplane in N-dimensional Euclidean space is a convex set. We will denote the scalar product of two vectors, x and y, as:

$$x^T y = \sum_{n=1}^{N} x_n y_n \tag{2.4}$$

Let a be a given vector and let b be a real number. A hyperplane perpendicular to a is the set S of points x which satisfy $a^T x = b$. That is, a hyperplane is the set:

$$S = \{x = (x_1, \ldots, x_N) : \; a^T x = b\} \tag{2.5}$$

Let x and y be two points in S, so

$$a^T x = b \quad \text{and} \quad a^T y = b \tag{2.6}$$

Multiplying these relations by α and by $1 - \alpha$, respectively, one obtains:

$$\alpha a^T x = \alpha b \quad \text{and} \quad (1 - \alpha) a^T y = (1 - \alpha) b \tag{2.7}$$

Adding these expressions yields

$$a^T(\alpha x + (1 - \alpha) y) = b \tag{2.8}$$

which implies that $\alpha x + (1 - \alpha) y \in S$. Thus any average of elements of S belongs to S. Hence S is convex.

2.2.3 Convexity in Function Spaces

The algebraic concept of convexity can be applied to sets of objects other than points in Euclidean space. For example, consider the set P of all density functions on the real line. These are non-negative functions $p(x)$ such that:

$$\int_{-\infty}^{\infty} p(x)dx = 1 \tag{2.9}$$

For any elements p and q in P and any $0 \leq \alpha \leq 1$, the convex combination of p and q is the function:

$$r(x) = \alpha p(x) + (1-\alpha)q(x) \tag{2.10}$$

$r(x)$ is clearly non-negative. Also,

$$\int_{-\infty}^{\infty} r(x)dx = \alpha \int_{-\infty}^{\infty} p(x)dx + (1-\alpha) \int_{-\infty}^{\infty} q(x)dx = 1 \tag{2.11}$$

Thus $r(x)$ is a density function and belongs to the set P. We see that any average of arbitrary elements of P belongs to P. Hence P is a convex set of functions.

Other convex sets of functions which we will encounter are:

$$A = \{f : |f(x)| \leq 1\} \tag{2.12}$$

$$B = \left\{f : \left|\frac{df}{dx}\right| \leq 1\right\} \tag{2.13}$$

$$C = \left\{f : \int_{-\infty}^{\infty} f^2(x)\,dx \leq 1\right\} \tag{2.14}$$

To prove that the set C is convex let g and h belong to C. Thus:

$$\int g^2(x)\,dx \leq 1 \quad \text{and} \quad \int h^2(x)\,dx \leq 1 \tag{2.15}$$

Let $u(x)$ be an arbitrary convex combination of g and h:

$$u(x) = \alpha g(x) + (1-\alpha)h(x) \tag{2.16}$$

We must show that the following integral does not exceed unity.

$$\int u^2(x)dx = \int \left[\alpha^2 g^2(x) + 2\alpha(1-\alpha)g(x)h(x) + (1-\alpha)^2 h^2(x)\right] dx \tag{2.17}$$

By the Schwarz inequality:

$$\int g(x)h(x)dx \leq \left[\int g^2(x)dx \int h^2(x)dx\right]^{1/2} \tag{2.18}$$

Hence, eq.(2.17) becomes:

$$\begin{aligned} \int u^2(x)dx \quad &\leq \quad \alpha^2 \int g^2(x)dx \\ &+ \quad 2\alpha(1-\alpha)\left[\int g^2(x)dx\right]^{1/2}\left[\int h^2(x)dx\right]^{1/2} \\ &+ \quad (1-\alpha)^2 \int h^2(x)\,dx \end{aligned} \tag{2.19}$$

None of the integrals on the righthand side of this equation exceed unity. Thus:

$$\int u^2(x)dx \leq \alpha^2 + 2\alpha(1-\alpha) + (1-\alpha)^2 = (\alpha + (1-\alpha))^2 = 1 \tag{2.20}$$

Hence, $u(x) \in C$ and C is a convex set of functions.

2.2.4 Set-Convexity and Function-Convexity

We have been discussing the convexity of sets of points or sets of functions. The concept of convexity is often applied to individual functions. While at first sight set-convexity and function-convexity may seem to be quite different, in fact they are related.

A usual definition of a convex function is the following. Let $f(x)$ be a function from E^N to E^1. Then f is a convex function if, for any convex coefficients $\alpha_1, \ldots, \alpha_N$,

$$f\left(\sum_{n=1}^{N} \alpha_n x_n\right) \leq \sum_{n=1}^{N} \alpha_n f(x_n) \tag{2.21}$$

For example, $f(x) = x^2$ and $f(x) = \mathrm{e}^x$ are convex functions, according to this definition (see Fig. 2.4.)

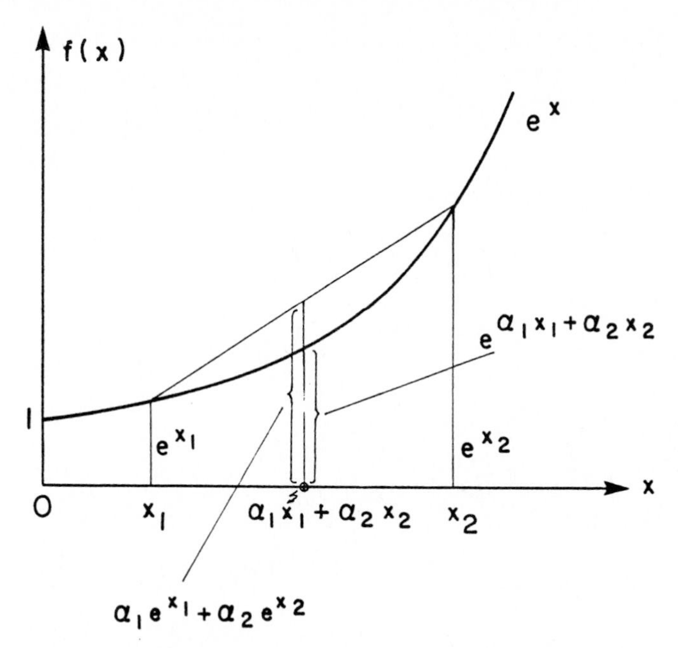

Figure 2.4: A convex function.

It can be shown that this definition is equivalent to the following definition in terms of convex sets. Let $g(x)$ be a function from a domain D in E^N to E^1. Define the *epigraph* of g as the set:

$$EG(g) = \{(x, y) : \ x \in D, y \in E^1, y \geq g(x)\} \tag{2.22}$$

$EG(g)$ is a set in E^{N+1} and contains all the points on or above the surface defined by $g(x)$. For example, see Fig. 2.5.

One can show (Rockafellar, 1970, section 4) that a function is convex according to eq.(2.21) if and only if its epigraph is a convex set.

2.3 The Structure of Convex Sets

2.3.1 Extreme Points and Convex Hulls

Consider the following set in E^2. See Fig. 2.6.

$$S = \{(x, y) : \ |x| \leq 1 \ , \ |y| \leq 1\} \tag{2.23}$$

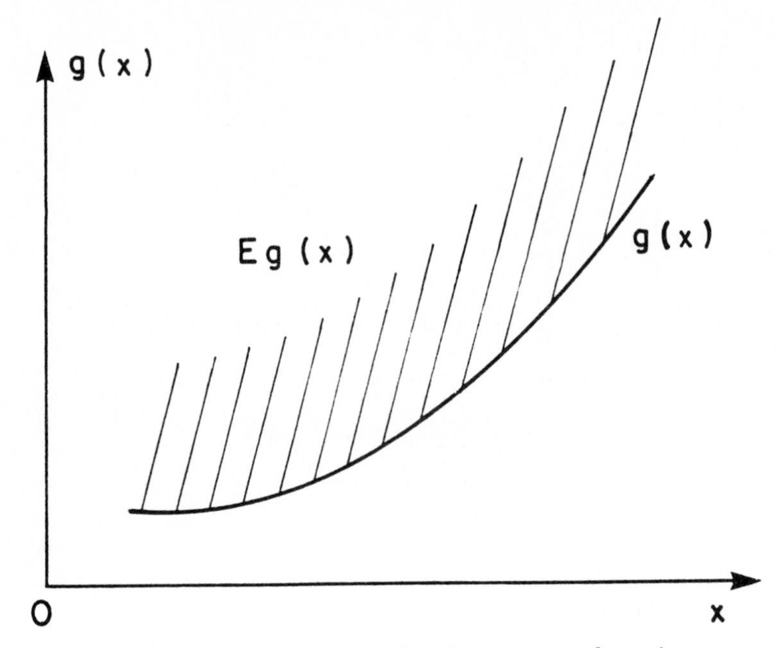

Figure 2.5: Epigraph of a convex function.

This is clearly a convex set. In fact any point in S can be expressed as a convex combination of the four points (1,1), $(1,-1)$ $(-1,-1)$ and $(-1,1)$. This is illustrated in Fig. 2.7, where an arbitrary point, P, is the convex combination of the points Q and (1,1), while Q is an average of $(-1,1)$ and $(-1,-1)$.

The *extreme points* of a set are those elements of the set which cannot be expressed as a convex combination of other elements in the set.

Let E be the set containing the four "corners" of S :

$$E = \{(1,1),\ (1,-1),\ (-1,-1),\ (-1,1)\} \tag{2.24}$$

E contains the extreme points of the set S.

The *convex hull* of a set A is the intersection of all convex sets containing A. The convex hull of A is denoted ch(A). Thus, loosely speaking, ch(A) is the "smallest" convex set containing A. See Fig. 2.8, where D, C and B are convex sets containing the set E. A non-convex set and its convex hull are shown in Fig. 2.9.

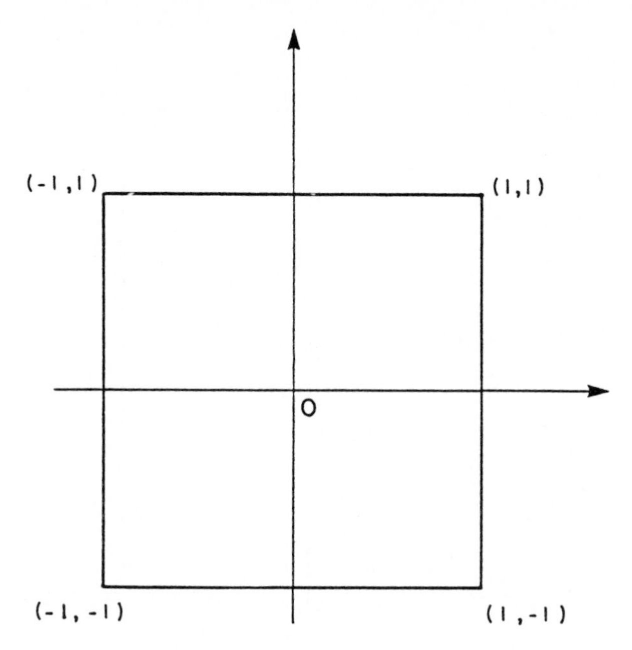

Figure 2.6: A convex set in E^2.

As a different example consider the spherical shell:

$$C = \left\{ x : \sum_{1}^{N} x_n^2 = 1 \right\} \tag{2.25}$$

The convex hull of C is the solid ball:

$$\text{ch}(C) = B = \left\{ x : \sum_{1}^{N} x_n^2 \leq 1 \right\} \tag{2.26}$$

The ideas of extreme points and convex hulls are connected by the following theorem.

Theorem 2.1. (Balakrishnan, 1981, p40). A closed and bounded[2] convex set in a Euclidean space is the convex hull of its extreme points.

The definition of the convex hull as an infinite intersection of convex sets does not reveal the really simple nature of the idea, which is expressed in the following theorem. We first require a definition.

---

[2]This theorem can be generalized to compact sets in a Hilbert space.

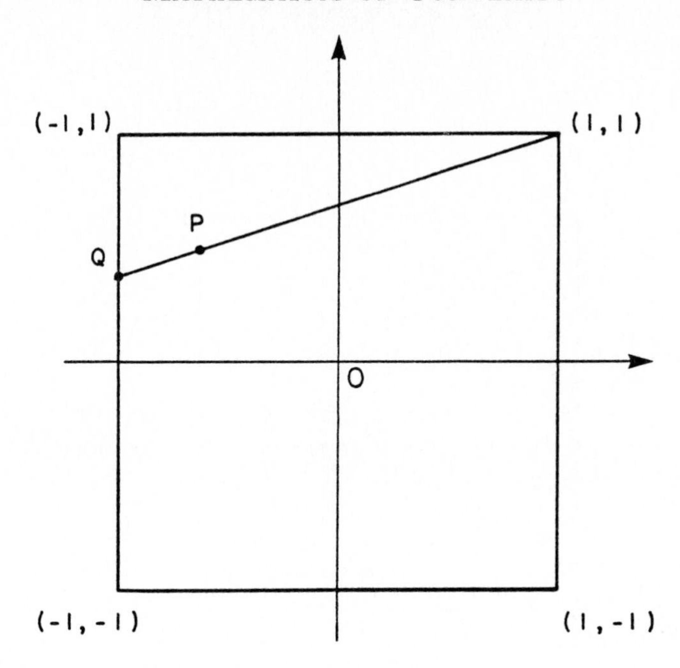

Figure 2.7: P is a convex combination of extreme points of S.

The *convex span* of a finite set of points is the set of all convex combinations of the elements in the set.

Theorem 2.2. (Kelly and Weiss, 1979, p 200). The convex hull of any set S is the union of the convex spans of all finite subsets of S.

Combining the last two theorems we see that a compact convex set is the set of all convex combinations of the extreme points of the set.

In fact, for sets in Euclidean space the situation is even simpler, as shown by the following theorem of Caratheodory.

Theorem 2.3. (Kelly and Weiss, 1979, p 201). If S is a set in E^N, then each point in ch(S) is a convex combination of $N+1$ or fewer points in S.

For example, examine the convex decomposition of the point P in Fig. 2.7.

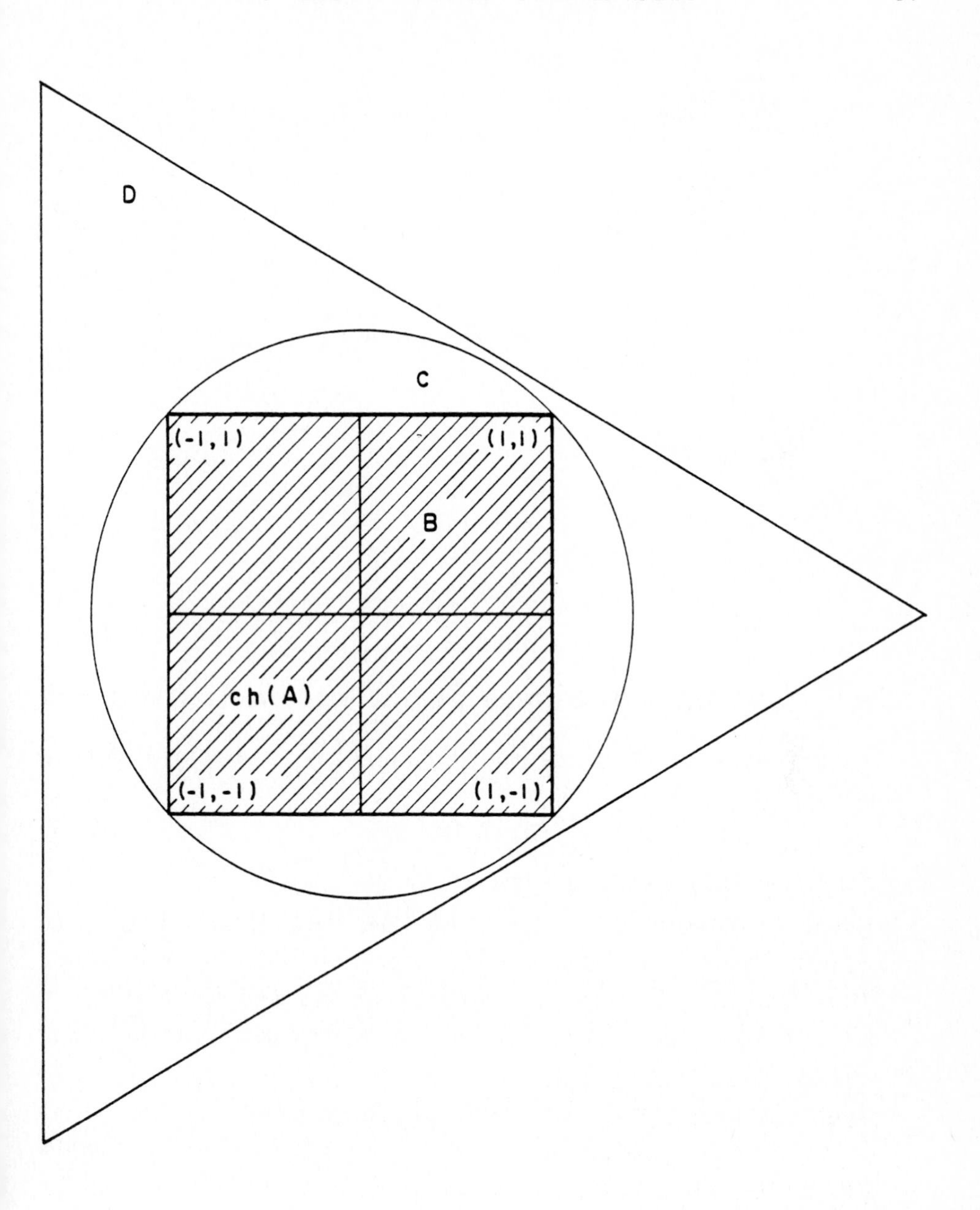

Figure 2.8: Convex sets containing the set E.

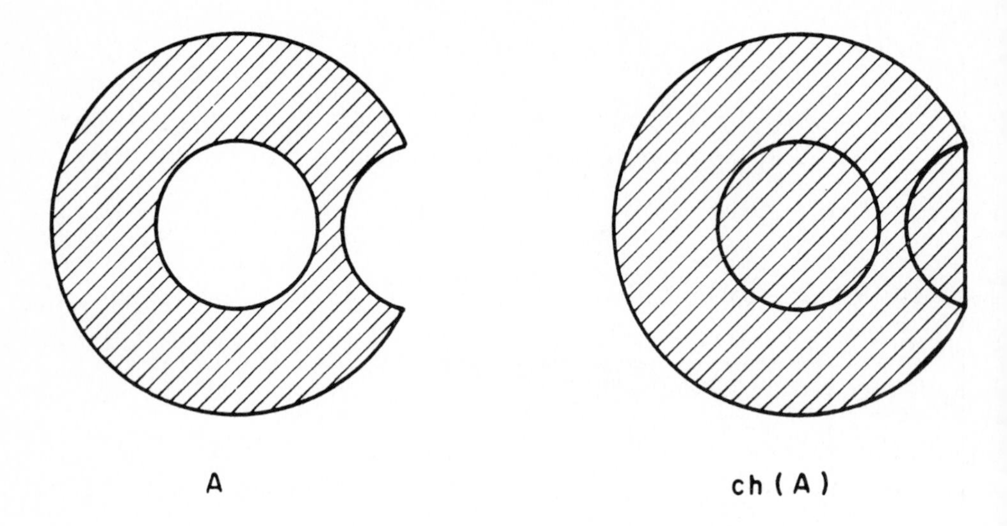

Figure 2.9: Convex hull of the non-convex set A.

2.3.2 Extrema of Linear Functions on Convex Sets

An affine function which maps from E^M to E^N is a matrix transformation of the form:

$$f(x) = Ax + v \tag{2.27}$$

where A is an $N \times M$ matrix and v is an N-vector.

Linear transformations between Euclidean spaces are obtained as the special case $v = 0$. Conversely, affine functions are translations of linear functions. In many applications we will seek the extrema of a linear or affine function on a compact, convex set. The following theorem is very useful.

Theorem 2.4. (Kelly and Weiss, 1979, p 209). If f is an affine function and S is a compact set, then f assumes the same minimum and maximum values on S and on ch(S).

Example. Let S be the rectangular region defined in eq.(2.23) and let E be the set of extreme points of S defined in eq.(2.24). Let f be the following linear function:

$$f(x, y) = x + y \tag{2.28}$$

We seek the maximum of f on S:

$$\max_{x,y\in S} f(x,y) = \max_{x,y\in E} f(x,y) \tag{2.29}$$

It is an elementary matter to find the maximum of $f(x,y)$ on the set E, which contains only four points. The maximum occurs for $(x,y) = (1,1)$, so:

$$\max_{x,y\in S} f(x,y) = 2 \tag{2.30}$$

Example. Consider the spherical shell and solid ball defined in eqs.(2.25) and (2.26). Let f be:

$$f(x) = a^T x \tag{2.31}$$

f is a linear function and we seek the extrema of f on B. Since f is a linear function we can seek the maximum on the set C of extreme points. This problem is readily solved with the method of Lagrange multipliers. The requirement that x belong to C can be expressed as:

$$x^T x - 1 = 0 \tag{2.32}$$

Define the objective function J as:

$$J = a^T x + \lambda(x^T x - 1) \tag{2.33}$$

A necessary condition for an extremum of J is:

$$0 = \frac{\partial J}{\partial x} = a + 2\lambda x \tag{2.34}$$

Hence the maximizing x is

$$x = -\frac{1}{2\lambda} a \tag{2.35}$$

This value of x must satisfy the constraint, eq.(2.32). Thus the Lagrange multiplier, λ, must assume one of the values:

$$\lambda = \pm\frac{1}{2}\sqrt{a^T a} \tag{2.36}$$

Finally, combining eqs.(2.35) and (2.36) we find the maximum value of $f(x)$ to be:

$$\max_{x\in B} f(x) = \sqrt{a^T a} \tag{2.37}$$

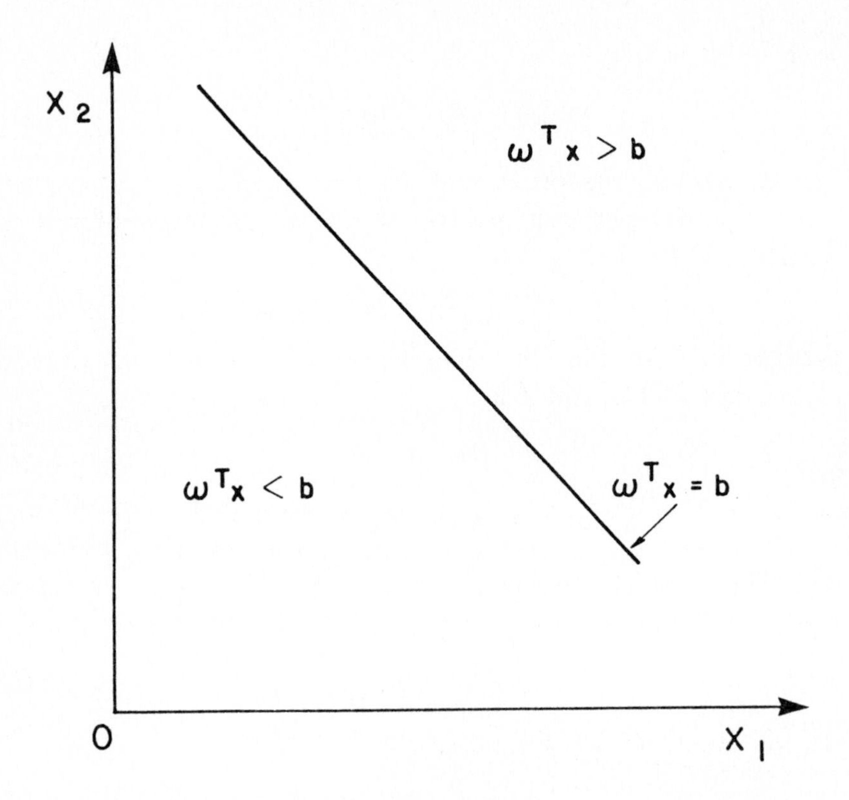

Figure 2.10: Half-spaces defined by a hyperplane.

2.3.3 Hyperplane Separation of Convex Sets

The algorithmic determination of the disjointness of convex sets arises in many applications, and depends on exploiting the unique structural properties of these sets. In this section we discuss the concept of hyperplane separation.

Let ω be a vector in E^N, so that $\omega^T x$ is a linear function from E^N to E^1. For any constant, b, the set of points x for which

$$\omega^T x = b \tag{2.38}$$

defines a hyperplane, P, in E^N, as in Fig. 2.10. The points x for which

$$\omega^T x < b \tag{2.39}$$

are those points on one side of the hyperplane P. Those points x for

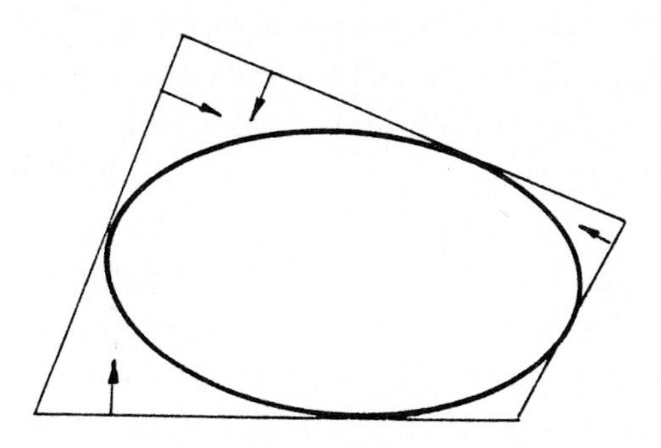

Figure 2.11: Half-spaces delimiting a convex set.

which

$$\omega^T x > b \tag{2.40}$$

are the points on the other side of P. Thus the plane $\omega^T x = b$ divides space into two half-spaces whose members are characterized by the sign of $\omega^T x$.

A closed half-space is the set of points in and on one side of a hyperplane. Closed half-spaces bear a special relation to closed convex sets, as expressed in the following theorem.

Theorem 2.5. (Rockafellar, (1970), p 99). A closed convex set is the intersection of all the closed half-spaces which contain it. (See Fig. 2.11).

These considerations give rise to the concept of *hyperplane separation* of convex sets, which is expressed in the following theorem.

Theorem 2.6. (Compare: Kelly and Weiss, 1979, p 145 or Stoer and Witzgall, 1970, p 98). Let C and F be non-empty, closed convex sets in E^N, and let one of them be bounded. C and F are disjoint if and only if there exists a hyperplane P such that C is in one half-space of P and F is in the other half-space.

This theorem can be expressed algebraically as follows:

$$C \cap F = \emptyset \tag{2.41}$$

if and only if there exists a real vector ω such that:

$$\max_{c \in C} \omega^T c < \min_{f \in F} \omega^T f \tag{2.42}$$

This relation forms the basis for an algorithmic determination of disjointness. The algorithm searches for a vector ω which satisfies relation (2.42). If such a vector is found, then the disjointness is established. If no such vector exists, then the sets intersect. Furthermore, the search for the extrema can be simplified by exploiting the fact that $\omega^T c$ and $\omega^T f$ are linear functions: The extrema of a linear functional on a compact convex set S occur on the extreme points of S (see Theorem 2.4). If C and F are the convex hulls of D and G respectively, then the extrema in (2.42) can be sought on D and G. Thus a necessary and sufficient condition for the disjointness of C and F is the existence of a vector ω such that:

$$\max_{d \in D} \omega^T d \; < \; \min_{g \in G} \omega^T g \tag{2.43}$$

2.4 Convex Models

In formulating a set-theoretic representation of uncertainty one typically begins by defining a convex set, R, of allowed functions. In order to easily maximize linear functions on this set of allowed functions one needs to know a set, E, whose convex hull is R. This pair of sets, E and R, is called a *convex model.* In the following examples we will present a number of commonly encountered convex models.

The set E must contain the extreme points of R, but it is not necessary that E contain only extreme points. In fact, it is sometimes convenient to allow E to be somewhat larger. Thus let us think of the set E as containing functions which are *fundamental* to the functions in the set R, where these fundamental functions need not be only extreme points of R. The process of finding a convex model is essentially the task of representing the set R of allowed functions by the set E of fundamental functions.

The representation of the functions in the set R in terms of the fundamental functions in the set E can be compared to representing a function as a series of orthogonal functions. The parallel:

orthogonal functions	$\longleftrightarrow$	fundamental functions
expansion coefficients	$\longleftrightarrow$	convex coefficients

An arbitrary function can be expanded in terms of orthogonal functions, where the contribution of each orthogonal function is expressed by an expansion coefficient. Similarly, elements of R can be represented as convex combinations of fundamental functions. The parallel should not be pushed too far, as important differences exist. However, it is useful to think of the set R of functions being decomposed into a simpler set E of extreme-point functions. The sets R and E are equivalent in the sense that the extrema of linear functionals on R and on E are equal.

We will now discuss several convex models.

Density Functions. The allowed functions are defined on an interval, J, and constrained to be non-negative and of constant integral value. The prototype is the set of density functions of unit mass:

$$R_{df} = \left\{ f : \ f(x) \geq 0 \ \text{ and } \ \int_J f(x)dx = 1 \right\} \tag{2.44}$$

The extreme points of this set are the Dirac δ-functions whose singularities fall in J.

$$E_{df} = \{ f : \ f(x) = \delta(x - \xi) \ \text{ for } \ \xi \in J \} \tag{2.45}$$

To see that every function in R_{df} is in fact the weighted average of elements in E_{df} we need only employ the basic definition of the Dirac δ-function:

$$f(x) = \int_J f(y)\delta(y - x)dy \tag{2.46}$$

Since $f(y)$ is itself a density function of unit mass on the domain J we see that $f(x)$ is a weighted average of elements of E_{df}.

Monotonic Functions. The set of real functions which increase monotonically from 0 to 1 on the interval [0,1] is:

$$R_{mf} = \{ f : \ f(0) = 0, \ f(1) = 1, \ f(x) \leq f(y) \ \text{ for } \ 0 \leq x \leq y \leq 1 \} \tag{2.47}$$

We will provide a heuristic derivation of the set of extreme points of R_{mf}. Let us note that each element of R_{mf} is the integral of an element of the set of density functions on the unit interval. We know that the extreme points of R_{df} are the Dirac δ-functions. Suppose that f is an extreme point of R_{mf} but not an integral of an extreme point of R_{df}. That is:

$$f(x) \in E_{mf} \quad \text{and} \quad \frac{df}{dx} \notin E_{df} \tag{2.48}$$

This means that there are distinct extreme points $g_1, g_2, \ldots,$ in E_{df} such that:

$$\frac{df}{dx} = \sum_i \alpha_i g_i(x) \tag{2.49}$$

where $\alpha_1, \alpha_2, \ldots$ are convex coefficients. This implies that:

$$f(x) = \sum_i \alpha_i \int_0^x g_i(\xi) d\xi \tag{2.50}$$

The convex combination on the righthand side belongs to R_{mf}. However, because f is an extreme point of R_{mf}, the functions $u_i(x) = \int_0^x g_i(\xi) d\xi$ must be identical:

$$\int_0^x g_i(\xi) d\xi = \int_0^x g_j(\xi) d\xi \tag{2.51}$$

for all $x \in [0, 1]$. This implies that:

$$g_i(x) = g_j(x) \tag{2.52}$$

which is a contradiction. Thus the supposition, eq.(2.48), is false. Consequently the extreme points of R_{mf} are integrals of the extreme points of R_{df}. Thus E_{mf}, the extreme points of R_{mf}, contains the unit step functions whose discontinuities occur anywhere in the unit interval. To define the elements of R_{mf} we need to define the *characteristic function*. Let V be a subset of E^N. The characteristic function $K_V(x) = 1$ if x belongs to V, and equals zero otherwise. Thus E_{mf} is

$$E_{mf} = \{f : \ f(x) = K_{[0,\xi]}(x) \quad \text{for} \quad 0 \le \xi \le 1\} \tag{2.53}$$

Uniformly Bounded Functions. Let R_{ubf} be the set of functions, defined on a domain J in E^N, which assume any values between $+1$ and -1. That is,

$$R_{ubf} = \{f : \ | f(x) | \le 1\} \tag{2.54}$$

The extreme points of R_{ubf} are the functions which assume only the values $+1$ and -1, and which switch back and forth between $+1$ and -1 at arbitrary points in J. Such functions are well known in control theory and are sometimes referred to as "bang-bang" or switching functions. The extreme points of R_{ubf} are conveniently represented using the characteristic function. Let U and V be disjoint subsets of E^N which just cover J. That is, the intersection of U and V is empty and the union of U and V equals J :

$$U \cap V = \emptyset \quad \text{and} \quad U \cup V = J \tag{2.55}$$

Consider the function:

$$f(x) = K_U(x) - K_V(x) \tag{2.56}$$

where U and V satisfy eq. (2.55). $f(x)$ is defined on J and assumes only the values $+1$ and -1. The extreme points of R_{ubf} are the functions f in eq. (2.56), for any choice of U and V satisfying the relations in (2.55). That is, the set of extreme points is:

$$E_{ubf} = \{f : \ f(x) = K_U(x) - K_V(x) \ \text{ for } \ U \cup V = J, U \cap V = \emptyset\} \tag{2.57}$$

A simple geometrical argument will make the choice of these switching functions plausible as the extreme points of R_{ubf}. The elements of R_{ubf} vary arbitrarily between $+1$ and -1. Let us consider a function $\varphi(x)$ which is bounded by ± 1 (and hence belongs to R_{ubf}) but whose value changes only N times in J. Thus $\varphi(x)$ is constant throughout each of N disjoint subsets of J. As N increases, $\varphi(x)$ can be chosen to approximate any element of R_{ubf} with arbitrary accuracy. It will suffice for our heuristic argument to show that any step-like function such as $\varphi(x)$ can be represented as a convex combination of the switching functions in E_{ubf}.

First consider the case $N = 1$, for which $\varphi(x)$ is constant throughout J. Let $f_1 = -1$ and $f_2 = +1$, as in Fig. (2.12). φ is obviously the average of f_1 and f_2 :

$$\varphi = \alpha f_1 + (1 - \alpha) f_2 \tag{2.58}$$

where $\alpha = (1 + \varphi)/2$.

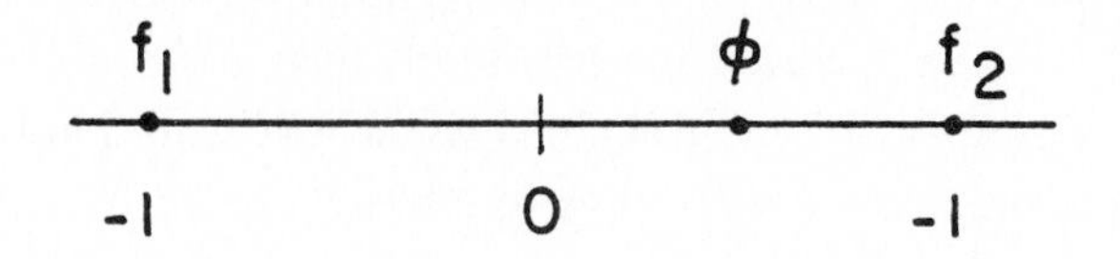

Figure 2.12: φ falls between f_1 and f_2 for $N = 1$.

Consider now the case $N = 2$. $\varphi(x)$ assumes only one of two values as x varies over J :

$$\varphi(x) = \begin{cases} \varphi_1 & , \quad x \in J_1 \\ \varphi_2 & , \quad x \in J_2 \end{cases} \tag{2.59}$$

where J_1 and J_2 are disjoint and cover J. We may think of φ as a vector: $\varphi = (\varphi_1, \varphi_2)$, where $\varphi(x) = \varphi_1$ while x is in J_1 and $\varphi(x) = \varphi_2$ while x is in J_2. Likewise, consider four switching functions: $f_1(x), \ldots, f_4(x)$, which change their values when $\varphi(x)$ does, as x varies on J. Let these four functions be represented by the vectors $(1,1)$, $(1,-1)$, $(-1,-1)$ and $(-1,1)$. Thus, for instance, $f_2(x) = 1$ while $\varphi(x) = \varphi_1$; $f_2(x) = -1$ while $\varphi(x) = \varphi_2$. The vectorial representations of φ and $f_1, \ldots, f_4$ are shown in Fig. (2.13). The point φ in this figure is contained in the convex region whose vertices are $f_1, \ldots, f_4$. Thus the point φ can be represented as a convex combination of the points $f_1, \ldots, f_4$. (Compare fig. (2.7)). It is evident that the same convex combination of the functions $f_1(x), \ldots, f_4(x)$ reproduces the function $\varphi(x)$.

Now consider the general case where $\varphi(x)$ assumes N values on J. Vectorially, φ can be represented as a point $\varphi = (\varphi_1, \ldots, \varphi_N)$ in E^N. Likewise, consider the 2^N vectors $f_n = (\overbrace{\pm 1, \ldots, \pm 1}^{N \text{ terms}})$, which define the vertices of a hypercube centered at the origin in E^N. The point φ is contained in this hypercube, and can be represented as a convex combination of the vertices. Consequently we see that $\varphi(x)$ can be represented as a convex combination of the 2^N distinct switching functions which jump between $+1$ and -1 as $\varphi(x)$ varies on J.

Our argument indicates (heuristically) that any function in R_{ubf} which changes its value N times in J can be represented as a convex combination of 2^N elements of E_{ubf}. By letting N increase without

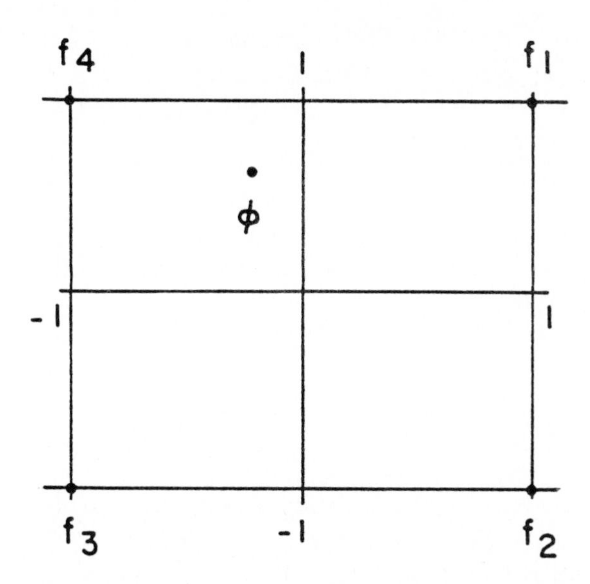

Figure 2.13: φ is contained in the square region defined by $f_1, \ldots, f_4$.

bound we see (intuitively, not rigorously) that arbitrary elements of R_{ubf} are convex combinations of elements of E_{ubf}. Thus R_{ubf} is the convex hull of E_{ubf}.

Envelope Bounded Functions. An immediate generalization of the uniform-bound model is to allow the bound to vary over the interval:

$$R_{ebf} = \{f : \; f_{\min}(x) \le f(x) \le f_{\max}(x)\} \tag{2.60}$$

where $f_{\min}(x)$ and $f_{\max}(x)$ are specified functions. The extreme points of R_{ebf} are:

$$\begin{aligned} E_{ebf} \;\; = \;\; & \{f : f(x) = K_V(x) f_{\max}(x) + K_U(x) f_{\min}(x) \\ & \text{for} \;\; U \cup V = J, U \cap V = \emptyset\} \end{aligned} \tag{2.61}$$

Functions With Bounded Derivatives. The uniformly bounded functions are constrained in their magnitude but not in their slope or curvature. It is sometimes useful to consider a set of functions whose derivatives are uniformly bounded. Consider, for example, the following set of functions with bounded first derivatives:

$$R_{bfd} = \left\{ f : \; \left| \frac{df}{dx} \right| \le 1 \right\} \tag{2.62}$$

Functions of Bounded Energy. An important variation of the uniformly bounded functions are the functions of bounded energy:

$$R_{fbe} = \left\{ f : \int_J f^2(x)\, dx \leq 1 \right\} \tag{2.63}$$

We discussed the convexity of this set in section 2.2.3. The set of extreme points of R_{fbe} is:

$$E_{fbe} = \left\{ f : \int_J f^2(x)\, dx = 1 \right\} \tag{2.64}$$

This can be understood intuitively by analogy to the fact that a solid ball in E^N is the convex hull of the spherical shell of the same radius and center: The integral of f^2 in eq.(2.64) is the limit of a sum of squares.

Unimodal Functions. In some applications one encounters functions which have a single maximum at a fixed point in their domain. Such sets can be defined in many different ways. As an example of a specific set of unimodal functions, define the function

$$n(x; c, w) = \frac{1}{\cosh\left(\frac{x-c}{w}\right)} \tag{2.65}$$

Thus $n(x; c, w)$ is a unimodal function centered at $x = c$ and with a width determined by the value of w. Consider the following set of extreme points:

$$E_{um} = \{ f : \; f(x) = n(x; c, w) \;\; \text{for} \;\; w_{\min} \leq w \leq w_{\max} \} \tag{2.66}$$

where c, $w_{\min}$ and $w_{\max}$ are specified. The convex hull of E_{um} contains functions which are all unimodal with their maxima at $x = c$. Note that for the first time we have defined a convex model by first specifying the extreme points.

Frequency-Bounded Functions. In some problems the allowed functions are specified in terms of constraints on their expansion coefficients with respect to a complete set of orthogonal functions, $\varphi_1, \varphi_2, \ldots$.

It is useful to formulate the convex model in the same manner. For example, suppose the functions are specified in terms of the first N coefficients, $c_1, \ldots, c_N$ in the expansion in $\varphi_1, \varphi_2, \ldots$. Let A be a convex set of N-tuples and define the set of allowed functions as:

$$R_{fbf} = \left\{ f : \; f = \sum_{n=1}^{N} c_n \varphi_n(x) \;\; \text{for} \;\; c \in A \right\} \tag{2.67}$$

For example, A may be the solid ellipsoid:

$$A = \left\{ c = (c_1, \ldots, c_N) : \; \sum_{n=1}^{N} \left(\frac{c_n}{\omega_n} \right)^2 \; \leq \; 1 \right\} \tag{2.68}$$

Thus A is the convex hull of the ellipsoidal shell:

$$B = \left\{ c = (c_1, \ldots, c_N) : \; \sum_{n=1}^{N} \left(\frac{c_n}{\omega_n} \right)^2 = 1 \right\} \tag{2.69}$$

R_{fbf} is the convex hull of:

$$E_{fbf} = \left\{ f : \; f = \sum_{n=1}^{N} c_n \varphi_n(x) \;\; \text{for} \;\; c \in B \right\} \tag{2.70}$$

Chapter 3

Uncertain Excitations

Parenthetically, let us remark that it is often impossible to assess the conceptual and analytical level of difficulty of a problem from its verbal formulation. Many questions easily posed in a few words from everyday usage escape even a formulation in precise terms.[1]

3.1 Introductory Examples

In this section we will discuss a number of elementary examples of additive excitations of linear systems. The purpose of the section is to illustrate the scope and method of the chapter as a whole. The discussion of this section will concentrate on the forced behaviour of a massless damped spring.

3.1.1 The Massless Damped Spring

Let $x(t)$ represent the displacement from equilibrium of the end of the spring, and let $f(t)$ represent the force applied to the spring. The equation of motion is

$$c\frac{dx(t)}{dt} + kx(t) = f(t) \tag{3.1}$$

[1]Bellman, 1961, p 204.

The general solution of eq. (3.1) is

$$x(t) = x(0)\exp(-t/\tau) + \frac{1}{c}\int_0^t \exp\left(-\frac{t-s}{\tau}\right) f(s)\,ds \tag{3.2}$$

where $\tau = c/k$.

3.1.2 Excitation Sets

The function $f(t)$ belongs to a convex set of excitation functions. This set represents the uncertainty in the force profile which is applied to the spring, and is called the convex model. Two convex models will be considered in this section. In one model the excitation functions are uniformly bounded, so the set of allowed forcing functions is

$$F_{a,b} = \{f :\ a \le f(t) \le b\} \tag{3.3}$$

Let $0 \le a < b$. The second convex model supposes that the energy of excitation is bounded, so the set of allowed excitation functions is

$$F_E = \left\{ f :\ \int_0^\infty f^2(t)dt \le E \right\} \tag{3.4}$$

3.1.3 Maximum Responses

Let us now assume that $x(0) = 0$, and seek the maximum distension of the spring which can be attained for any allowed excitation. Consider first the uniformly bounded excitations. Denote the maximum distension as $\hat{x}_{ub}(t)$ which is

$$\hat{x}_{ub}(t) = \max_{f \in F_{a,b}} x(t) \tag{3.5}$$

$$= \max_{f \in F_{a,b}} \frac{1}{c}\int_0^t \exp\left(-\frac{t-s}{\tau}\right) f(s)\,ds \tag{3.6}$$

It is evident that the maximum is attained when $f(s) = b$. Hence we find that the greatest possible distension of the spring at time t is

$$\hat{x}_{ub}(t) = \frac{b\tau}{c}\left(1 - e^{-t/\tau}\right) \tag{3.7}$$

Let $\hat{x}_E(t)$ denote the greatest distension of the spring at time t subject to excitations of bounded energy. Thus

$$\hat{x}_E(t) = \max_{f \in F_E} \frac{1}{c} \int_0^t \exp\left(-\frac{t-s}{\tau}\right) f(s)\, ds \tag{3.8}$$

Because $x(t)$ is a linear function of the excitation, $f(t)$, and because the excitation set, F_E, is convex, the maximum in eq. (3.8) occurs on the set of extreme points of F_E. Thus the maximizing excitation will satisfy

$$\int_0^t f^2(s)\, ds = E \tag{3.9}$$

and $f(s) = 0$ for $s > t$.

The Cauchy-Schwarz inequality (Hardy *et al*, 1934, p 16) asserts that, for arbitrary functions u and v,

$$\left(\int_0^t u(s) v(s) ds \right)^2 \leq \int_0^t u^2(s)\, ds \int_0^t v(s)^2\, ds \tag{3.10}$$

with equality only if $u(s)$ and $v(s)$ are proportional. Thus the maximum in eq. (3.8) is attained if $f(s)$ is proportional to $\exp(-(t-s)/\tau)$, where the constant of proportionality assures that $f(s)$ satisfies eq. (3.9). The maximizing excitation profile on the interval $[0,t]$ is:

$$f(s) = \sqrt{\frac{2E}{\tau}} \left(1 - \mathrm{e}^{-2t/\tau}\right)^{-1/2} \exp\left(-\frac{t-s}{\tau}\right) \tag{3.11}$$

Combining this with eq. (3.8) one finds the greatest possible distension of the spring at time t to be

$$\hat{x}_E(t) = \frac{1}{c} \sqrt{\frac{\tau E}{2} \left(1 - \mathrm{e}^{-2t/\tau}\right)} \tag{3.12}$$

3.1.4 Measurement Optimization

In the previous examples we have determined various extremal properties of the responses to specified convex excitation sets. Problems of this sort arise in designing a system whose response must satisfy known constraints. In some situations, however, one wishes to characterize the excitation set itself by measuring the system response. It is then necessary to design the measurement so as to be able to differentiate between distinct excitation sets. Measurement optimization problems such as this are quite amenable to analysis, and have been extensively studied.

Let $F_{a,b}$ and $F_{\alpha,\beta}$ be two convex models for uniformly bounded excitation of the damped spring. Consider the measurement of the displacement of the spring at N instants, $t_1, \ldots, t_N$, where $0 \leq t_1 < \cdots < t_N \leq T$. Let $\xi_n = x(t_n)$ represent the measurement of the length of the spring at time t_n, and represent the vector of N measurements by $\xi = (\xi_1, ..., \xi_N)$. We wish to choose the least number of measurements and to specify their precise timing so that $F_{a,b}$ and $F_{\alpha,\beta}$ are distinguishable.

The set of all possible measurement vectors ξ in response to excitations from $F_{a,b}$ is the complete response set $C_{a,b}$:

$$C_{a,b} = \left\{ \xi : \ \xi_n = \frac{1}{c} \int_0^{t_n} f(s) \exp\left(-\frac{t_n - s}{\tau} \right) ds \quad , \quad f \in F_{a,b} \right\} \tag{3.13}$$

The complete response set $C_{\alpha,\beta}$ for excitations in $F_{\alpha,\beta}$ is similarly defined.

The excitation sets are convex and the measurement vector is an affine transformation of the excitation. Thus the response sets are convex. As discussed in section 2.3.3, a necessary and sufficient condition for disjointness of the response sets is the existence of a hyperplane which separates them. That is,

$$C_{a,b} \cap C_{\alpha,\beta} = \emptyset \tag{3.14}$$

if and only if there is a vector ω which satisfies

$$\max_{\xi \in C_{a,b}} \omega^T \xi \ < \ \min_{\zeta \in C_{\alpha,\beta}} \omega^T \zeta \tag{3.15}$$

We now proceed to develop expressions for the extrema in relation (3.15). Let $t_0 = 0$. It is evident that

$$\omega^T \xi \ = \ \sum_{n=1}^{N} \frac{\omega_n}{c} \int_0^{t_n} f(s) \exp\left(-\frac{t_n - s}{\tau} \right) ds \tag{3.16}$$

$$= \sum_{n=1}^{N} \frac{\omega_n}{c} \int_{t_{n-1}}^{t_n} f(s) \sum_{m=n}^{N} \exp\left(-\frac{t_m - s}{\tau}\right) ds \tag{3.17}$$

Let us define

$$\sigma_n = \sum_{m=n}^{N} \mathrm{e}^{-t_m/\tau} \quad , \quad n = 1, \ldots, N \tag{3.18}$$

Thus eq. (3.17) becomes

$$\omega^T \xi = \sum_{n=1}^{N} \frac{\sigma_n \omega_n}{c} \int_{t_{n-1}}^{t_n} f(s) \exp(s/\tau)\, ds \tag{3.19}$$

Clearly each integral in eq. (3.19) achieves its maximum on $F_{a,b}$ if $f(s)$ is chosen in the interval $[t_{n-1}, t_n]$ as

$$f(s) = \begin{cases} b & , \quad \omega_n \geq 0 \\ a & , \quad \omega_n < 0 \end{cases} \tag{3.20}$$

Let $H(x)$ equal unity if $x \geq 0$ and equal zero otherwise. The maximum of $\omega^T \xi$ becomes

$$\max_{\xi \in C_{a,b}} \omega^T \xi = \sum_{n=1}^{N} \frac{\sigma_n \omega_n \tau}{c} \left(\mathrm{e}^{t_n/\tau} - \mathrm{e}^{t_{n-1}/\tau}\right) (bH(\omega_n) + aH(-\omega_n)) \tag{3.21}$$

A similar argument demonstrates that the minimum of $\omega^T \zeta$ on $C_{\alpha,\beta}$ is

$$\min_{\zeta \in C_{\alpha,\beta}} \omega^T \zeta = \sum_{n=1}^{N} \frac{\sigma_n \omega_n \tau}{c} \left(\mathrm{e}^{t_n/\tau} - \mathrm{e}^{t_{n-1}/\tau}\right) (\alpha H(\omega_n) + \beta H(-\omega_n)) \tag{3.22}$$

Combining eqs. (3.21) and (3.22) with eq. (3.15) one finds that a necessary and sufficient condition for the disjointness of $C_{a,b}$ and $C_{\alpha,\beta}$ is the existence of a vector ω such that

$$0 < \sum_{n=1}^{N} \frac{\sigma_n \omega_n \tau}{c} \left(\mathrm{e}^{t_n/\tau} - \mathrm{e}^{t_{n-1}/\tau}\right) ((\alpha - b)H(\omega_n) + (\beta - a)H(-\omega_n)) \tag{3.23}$$

Various analytical and numerical techniques are available for searching for a vector ω satisfying this relation.

3.2 Vehicle Vibration

3.2.1 Introduction

A vehicle traveling along an irregular substrate experiences stresses and accelerations due to its reaction to the ground. If the profile of the substrate is precisely known, however complex it might be, then the vehicle response can be calculated in a straightforward manner either in closed form or numerically. In practice however, the irregularities of the substrate are often inaccurately known, so that conventional deterministic structural analysis is inapplicable. In most studies, therefore, probabilistic analysis has been adopted. Spectral analysis has been widely used to probabilistically model both the structure of an uncertain substrate and its influence on a moving vehicle.

Spectral analysis typically assumes that the vehicle moves with constant horizontal velocity and in continuous contact with a weakly homogeneous random substrate. This analysis has been pioneered by Kozin and Bogdanoff (1960) and by Robson (1963) and has been successfully studied by a number of other investigators. The most comprehensive early study, which included a probabilistic description of the road roughness, random vibration analysis of the vehicle and effects of the vibrations on the payload, was conducted by Kozin, Bogdanoff and Cote (1965, 1966). The modern state of the art is discussed in the books of Bekker (1969), Nikolaenko (1967), and Robson *et al* (1971) and in the review articles by Schiehlen (1986a, 1986b, 1988), Newland (1986) and Robson (1984) (see also Sobczyk *et al,* 1976, 1977).

While spectral analysis is of proven power and versatility, some limitations can be discerned. In the study of vehicle vibrations the substrate is typically assumed to be described as a stationary Gaussian random process, and then spatial correlation functions are measured. The required data banks are not always available, and the assumption of stationarity and gaussianity is not universally valid.

From the perspective of convex modelling, the uncertainty in the

---

This section represents an extended version of material presented at the ASME Pressure Vessel and Piping Conference, Honolulu, Hawaii, June 1989. (Ben-Haim and Elishakoff, (1989b)). The authors are indebted to Osnat Katzanek and Dror Saddan for assistance in performing the numerical calculations.

substrate is described by a set of allowed substrate-profile functions. The definition of this set is either based on limited available information or, conversely, constitutes design specifications for the substrate. The analysis of the vehicle response consists in determining the range of variation of performance parameters, as the substrate profile varies over the set of allowed functions. This enables characterization of the uncertain vehicle motion as well as optimization of the vehicle design with respect to uncertainty in the substrate.

An important feature of the set-theoretical methodology is its adaptability to a wide range of substrate morphologies, and to the various forms in which partial information about the substrate is available. In sections 3.2.3 to 3.2.5 the substrate is assumed to be uniformly bounded; in section 3.2.6 uniform bounds on the variation in the slope of the substrate is incorporated; in section 3.2.7 the substrate is assumed to be a barrier or obstacle of uncertain shape. The vehicle model studied here is from Kozin and Bogdanoff (1960) and Newland (1986).

3.2.2 The Vehicle Model

The theory to be developed in this study will be applied to the vertical response of the center of mass of a two-wheeled trailer, as shown in Fig. 3.1 (Newland, 1986). The vertical displacement imposed by the substrate on the ith wheel is denoted $y_i(t)$. The forward velocity of the vehicle is constant in time. The equation of motion of the vertical displacement $x(t)$ of the center of mass is

$$M\frac{d^2x}{dt^2} + 2c\frac{dx}{dt} + 2kx = c\left(\frac{dy_1}{dt} + \frac{dy_2}{dt}\right) + k(y_1 + y_2) \tag{3.24}$$

where:

M = mass of the trailer body (mass of wheels and suspension are neglected).

c = damping coefficient of each suspension.

k = stiffness of each suspension.

The initial conditions are

$$x(0) = \frac{dx(0)}{dt} = 0 \tag{3.25}$$

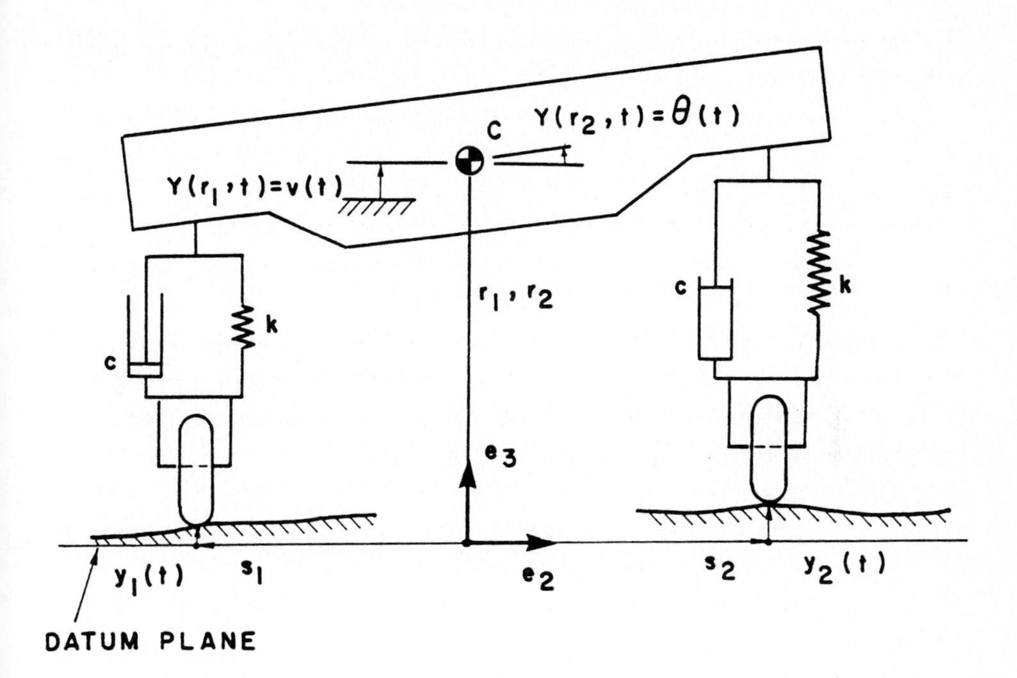

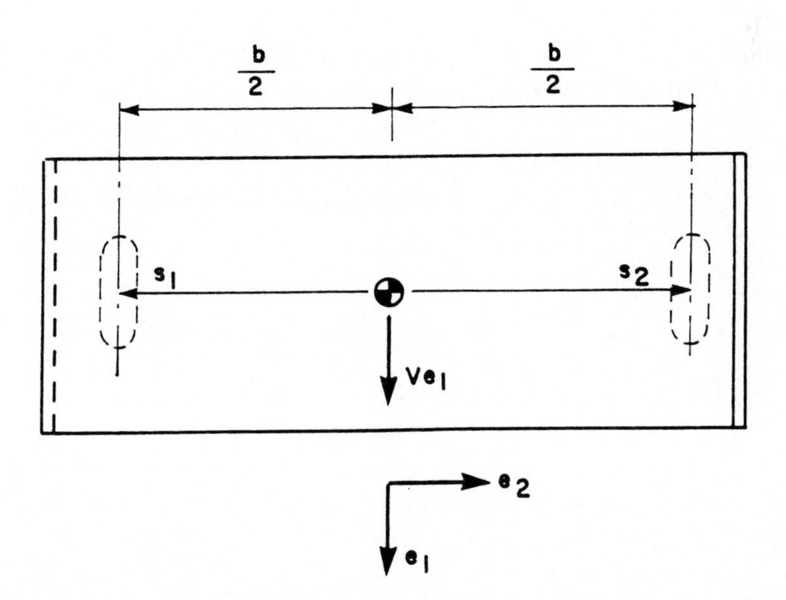

Figure 3.1: Two-wheeled trailer (after Newland, 1986). By permission of D.E. Newland and Elsevier Science Publishers, Amsterdam.

We assume also that $y_i(0) = y_i'(0) = 0$, where the prime denotes differentiation with respect to time.

For convenience let us define the following parameters:

$$\omega_o^2 = \frac{2k}{M} \quad , \quad \zeta\omega_o = \frac{c}{M} \tag{3.26}$$

$$\beta = \sqrt{1-\zeta^2} \quad , \quad \omega_o\beta = \omega_d \tag{3.27}$$

The values of these parameters, based on the data presented in Newland (1986), are: $\omega_o = 8.944\mathrm{s}^{-1}$, $\zeta = 0.1118$, $\beta = 0.9937$ and $\omega_d = 8.888\mathrm{s}^{-1}$.

The general solution of eq. (3.24) for $\zeta < 1$ is obtained, after partial integration to remove the derivatives of y_i, as

$$x(t) = \int_0^t [y_1(\tau) + y_2(\tau)]\varphi(t-\tau)d\tau \tag{3.28}$$

where

$$\varphi(t) = \frac{\omega_o}{\beta}\mathrm{e}^{-\zeta\omega_o t}\left[\left(\frac{1}{2} - \zeta^2\right)\sin\omega_d t + \zeta\beta\cos\omega_d t\right] \tag{3.29}$$

Likewise, the vertical velocity $v(t)$ and acceleration $a(t)$ of the center of mass are obtained by differentiation of $x(t)$, for $\zeta < 1$, as

$$v(t) = \zeta\omega_o[y_1(t) + y_2(t)] + \int_0^t [y_1(\tau) + y_2(\tau)]\psi(t-\tau)d\tau \tag{3.30}$$

$$\begin{aligned} a(t) &= \zeta\omega_o\left(\frac{dy_1}{dt} + \frac{dy_2}{dt}\right) + \omega_o^2\left(\frac{1}{2} - 2\zeta^2\right)[y_1(t) + y_2(t)] \\ &+ \int_0^t [y_1(\tau) + y_2(\tau)]\rho(t-\tau)d\tau \end{aligned} \tag{3.31}$$

where

$$\psi(t) = \frac{\omega_o^2}{\beta}\mathrm{e}^{-\zeta\omega_o t}\left[\zeta\left(\zeta^2 - \beta^2 - \frac{1}{2}\right)\sin\omega_d t + \beta\left(\frac{1}{2} - 2\zeta^2\right)\cos\omega_d t\right] \tag{3.32}$$

$$\rho(t) = \frac{\omega_o^3}{\beta} e^{-\zeta\omega_o t} \left[\left(3\beta^2\zeta^2 + \frac{1}{2}\left(\zeta^2 - \beta^2\right) - \zeta^4 \right) \sin\omega_d t \right.$$
$$\left. + \; \zeta\beta\left(3\zeta^2 - \beta^2 - 1\right)\cos\omega_d t \right] \tag{3.33}$$

Note that $\psi(t) = \varphi'(t)$ and $\rho(t) = \psi'(t)$.

3.2.3 Uniformly Bounded Substrate Profiles

The human response to vibration is an important parameter to be considered in designing vehicles for traversing rough terrain. Extensive data (Harris, 1987) have been accumulated for evaluating the effect of vibration on humans. For example, data are available which indicate the average peak accelerations at which human subjects first perceive the vibration, then find it unpleasant and finally refuse to tolerate the vibration. Accepted standards establish quantitative guidelines for vibration exposure in vehicles. These guidelines limit the duration of exposure to vibrational acceleration for preservation of health and safety ("exposure limits"); for preservation of work efficiency ("decreased proficiency boundaries"); and for preservation of comfort ("reduced comfort boundary"). These guidelines are based on experimental measurements with human subjects under idealized conditions of constant frequency and uniform vibrational modes. The guidelines express only approximate limits for safety, work-efficiency and comfort since laboratory tests cannot reproduce the complete range of subject-vehicle-environment interactions combined with the inevitable physiological and psychological variations of the subjects. Nevertheless, the standards provide benchmark data with respect to which vehicle performance on rough terrain can be evaluated.

In this section the peak vertical acceleration of the vehicle described in section 3.2.2 is evaluated, for motion on a rough but uniformly bounded substrate. This analysis, together with the standards for vibrational acceleration, provides the basis for vehicle-design to assure satisfactory performance with respect to vibrational acceleration when traversing a rough surface. The following section discusses the duration of acceleration excursions.

An actual substrate is represented by a function which describes the vertical displacement of the substrate with respect to the nominal profile, at each point on the plane. The uncertainty in the substrate profile is modelled by specifying a set of allowed displacement functions. This set approximately represents the ensemble of all profiles which can be encountered in practice.

The uniform-bound model of uncertainty in the substrate profile asserts that an actual substrate can deviate above or below the nominal, flat, horizontal profile at any point by no more than a specified distance. This implies that there is no relation between substrate displacements at distinct points of the profile.

Consequently, the vertical displacements imposed on the vehicle [$y_1(t)$ and $y_2(t)$ in eq. (3.24)] each belong to the following set of uniformly bounded functions:

$$Y(\hat{y}) = \{y : \ |y(t)| \leq \hat{y} \quad , \quad 0 \leq t < \infty\} \tag{3.34}$$

This means that the actual substrate at any point can deviate from the nominally flat substrate by as much as $\pm\hat{y}$. The value of $\hat{y}$ either represents available information about a real substrate or, conversely, is a design parameter to be determined in specifying the substrate.

It will be noticed that $Y(\hat{y})$ is a convex set of functions. That means that, for any functions $u(t)$ and $w(t)$ in $Y(\hat{y})$ and any $0 \leq \alpha \leq 1$, the function $\alpha u(t) + (1-\alpha)w(t)$ also belongs to the set $Y(\hat{y})$. In other words, any convex combination of allowed substrate profiles is also an allowed profile.

3.2.4 Extremal Responses On Uniformly Bounded Substrates

The first performance parameters to be examined in the uniform-bound model are the maximum vertical vehicle displacement, velocity and acceleration. Examination of eq. (3.28) shows that the vertical displacement of the center of mass of the vehicle at time t attains a maximum if $y_1(\tau)$ and $y_2(\tau)$ are of the same sign as $\varphi(t-\tau)$ and equal in magnitude to $\hat{y}$. Thus the extremal displacement due to uniformly bounded terrain, $\hat{x}_{ub}(t)$, is (for $\zeta < 1$)

$$\hat{x}_{ub}(t) = 2\hat{y}\int_0^t |\varphi(\tau)|d\tau \tag{3.35}$$

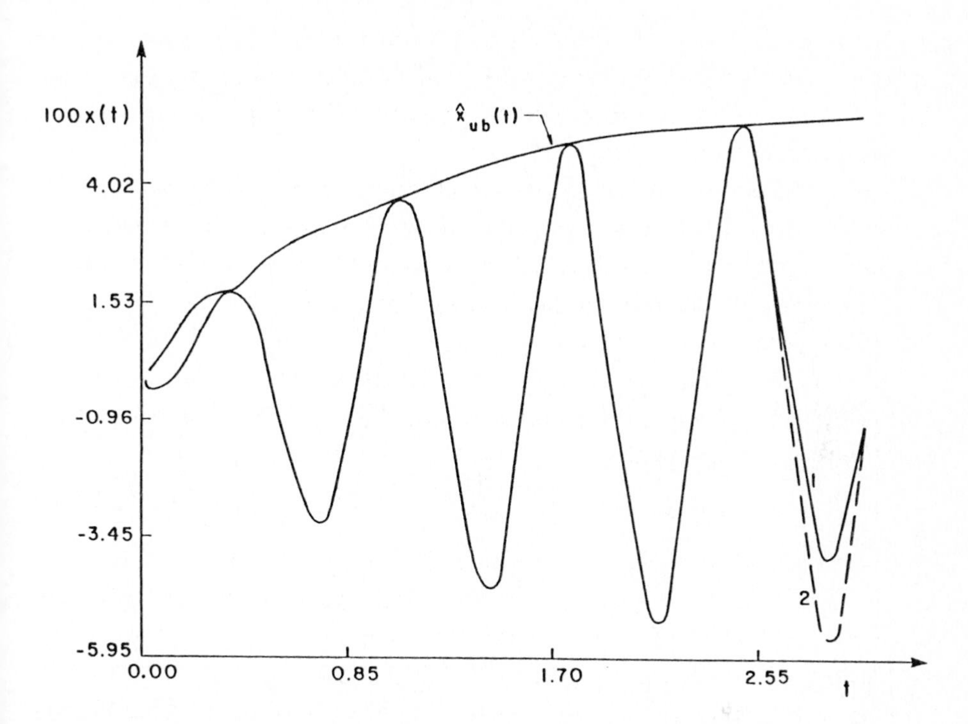

Figure 3.2: Envelope of maximum vertical displacement of center of mass (upper curve), and displacement history reaching the envelope at several instants (lower curve).

$\hat{x}_{ub}(t)$ represents the least upper bound at time t of all possible displacement histories. In other words, $\hat{x}_{ub}(t)$ is the upper envelope of the displacement in all possible histories, but is not itself a displacement history.

A similar expression can be obtained from eq. (3.30) for the maximum vertical velocity. The least upper bound at time $t > 0$ of the vertical velocity of the center of mass is $\hat{v}_{ub}(t)$

$$\hat{v}_{ub}(t) = 2\hat{y}\zeta\omega_o + 2\hat{y}\int_0^t |\psi(\tau)|d\tau \tag{3.36}$$

The vehicle responses are maximized by a "bang-bang" road profile which switches back and forth between its extremal values. This results

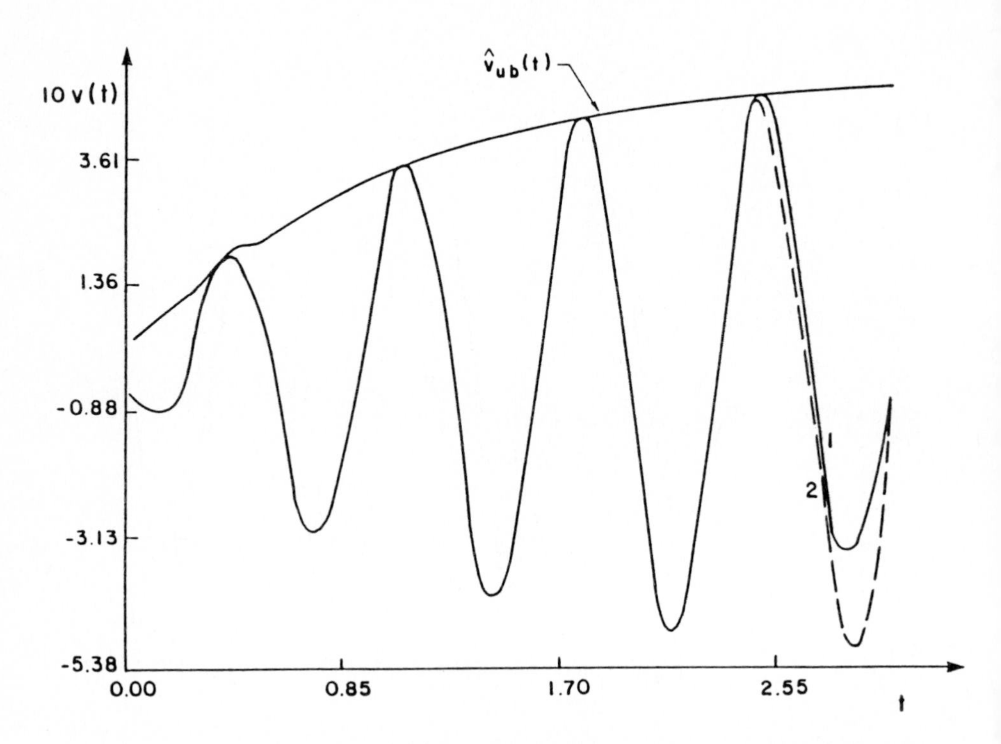

Figure 3.3: Envelope of maximum vertical velocity of center of mass (upper curve), and velocity history reaching the envelope at several instants (lower curve).

in impulse forces applied to the center of mass at the instants of road-profile discontinuity, as indicated by the derivatives on the righthand side of eq. (3.31). In addition, there act the step loads at these instants, as the last two terms of eq. (3.31) show. Aside from the impulse accelerations, the envelope of maximal vertical acceleration is found from eq. (3.31) to be

$$\hat{a}_{ub}(t) = 2\hat{y}\omega_o^2 \left| \frac{1}{2} - 2\zeta^2 \right| + 2\hat{y} \int_0^t |\rho(\tau)| d\tau \tag{3.37}$$

The "bang-bang" characteristic of the road profiles which maximize the responses is reminiscent of "bang-bang" control functions. These frequently arise when linear inequality constraints are placed on the

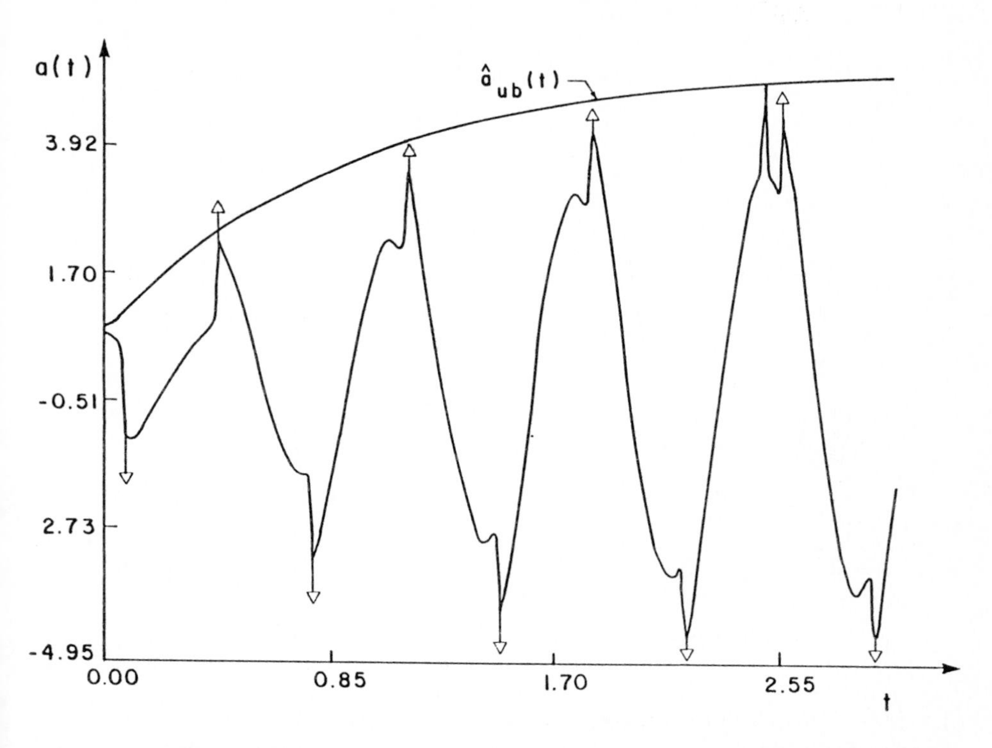

Figure 3.4: Envelope of maximum vertical acceleration of center of mass (upper curve), and acceleration history reaching the envelope at several instants (lower curve).

control variables (Bryson and Ho, 1975), or in minimal-time control when the system is "moved from one state to another in the shortest time by all times properly utilizing all available power" (Ogata, 1970).

Figs. 3.2 to 3.4 depict the envelopes of maximum vertical displacement, velocity and acceleration of the center of mass, respectively. The functions $\varphi(t), \psi(t)$ and $\rho(t)$ are shown in Fig. 3.5. The terrain profiles which maximize each of the responses for a preselected instant $t = t^*$ are periodic "bang-bang" profiles. The phase of the "bang-bang" profile $y(t)$ which maximizes the displacement at time t^* is such that $y(0)$ and $\varphi(t^*)$ are of the same sign, and such that the first switch in the profile occurs at the smallest non-negative value of τ for which $\varphi(t^* - \tau) = 0$. The period of the maximizing profile is $2\pi/\omega_d = 0.707$ sec. The phase and period of maximizing profiles for the velocity and acceleration are

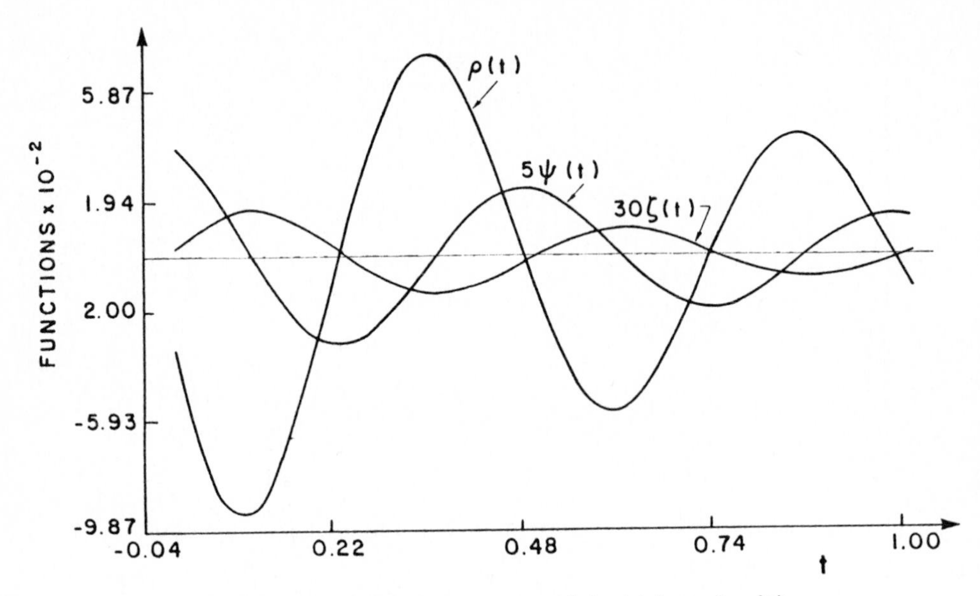

Figure 3.5: The functions $\varphi(t), \psi(t)$ and $\rho(t)$.

obtained by applying similar considerations to ψ and ρ, respectively.

Fig. 3.2 portrays, along with the envelope $\hat{x}_{ub}(t)$, a displacement history which reaches the envelope at $t^* = 2.5$ sec. The displacement history is continued after 2.5 sec for two cases: Curve 1 is associated with the vehicle continuing on a flat road with $y(t > 2.5) = 0$, whereas curve 2 is associated with the vehicle remaining on a "bang-bang" profile with the same period as the one which maximizes the displacement at 2.5 sec. Interestingly, there are a number of instants prior to 2.5 sec at which the displacement history $x(t)$ touches the envelope $\hat{x}(t)$, as seen in Fig. 3.2. This occurs whenever t^* is greater than the period of $\varphi(t)$.

Fig. 3.3 shows the envelope of maximal velocities, $\hat{v}_{ub}(t)$, together with a velocity history reaching a maximum at $t^* = 2.5$ sec. Two components in this velocity history can be discerned. One component is regularly *undulating,* as expected by the integral term in eq. (3.30). The second component consists of jumps in the velocity of magnitude $2\zeta\omega_0\hat{y} = 0.02\text{m/sec}$. These jumps, which are not visible in Fig. 3.3, occur at the instants of road-profile discontinuity. The history is continued beyond 2.5 sec on the same two profiles indicated in connection

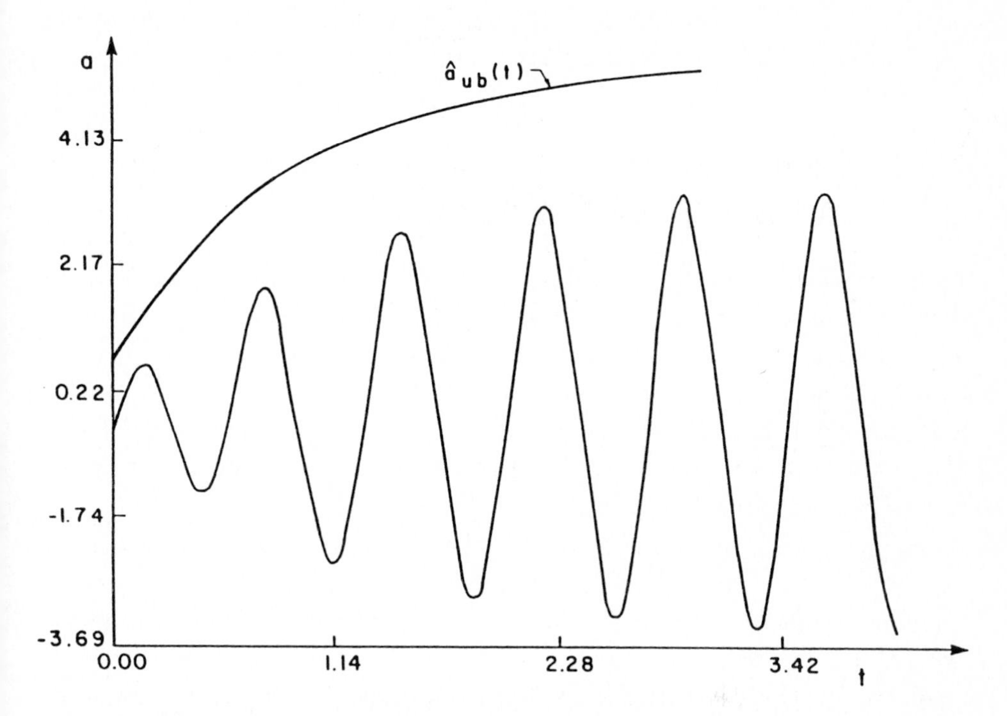

Figure 3.6: Envelope of maximum vertical acceleration of center of mass (upper curve), and acceleration history on a sinsusoidal road profile (lower curve).

with Fig. 3.2.

Fig. 3.4 shows the envelope of maximum acceleration, $\hat{a}_{ub}(t)$, with an acceleration history reaching the envelope at $t^* = 2.5$ sec. The uniform bound on the road-profile is $\hat{y} = \pm 0.01$m. Eq. (3.31) reveals that $a(t)$ contains derivatives of the "bang-bang" profile, i.e. Dirac delta functions, at the instants when the profile-discontinuities occur. These impulse accelerations are represented in Fig. 3.4 by arrows. In addition, the term $\omega_o^2(0.5 - 2\zeta^2)[y_1(t) + y_2(t)]$ indicates that the acceleration history contains jumps of amplitude 1.52 $\mathrm{m/s^2}$ at the instants when the profile-discontinuities occur. This is quite evident in Fig. 3.4.

Fig. 3.6 shows the acceleration history in response to a sinusoidal road with amplitude 0.01m and phase and period equal to the maximizing "bang-bang" profile. The maximizing acceleration envelope is reproduced here from Fig. 3.4. It is evident that the sinusoidal profile is significantly less severe than the "bang-bang" profile.

3.2.5 Duration of Acceleration Excursions on Uniformly Bounded Substrates

Accepted guidelines limiting the exposure to vibrational acceleration refer to vibration at a fixed frequency. In actual travel over an irregular substrate the vibrational acceleration is composed of a range of frequencies. However, the net effect of vertical vibrational acceleration is an up-down motion, regardless of the precise frequency composition of the acceleration waveform. In other words, the acceleration changes sign in a roughly periodic fashion.

A parameter which is related to the frequency-structure of the acceleration is the maximum duration for which the acceleration exceeds a given threshold. A rapidly oscillating vibrational acceleration maintains the acceleration above any given value for only a short duration. Conversely, slowly oscillating acceleration displays long durations of acceleration excursions above a given threshold. Evaluation of the greatest duration of vibrational acceleration above a given threshold enables partial assessment of the typical frequency of the vibrational acceleration. This is important since the standards for acceptable vibration vary with the frequency of the vibration.

The linearity and homogeneity of eq. (3.24), describing the vehicle vibration, result in an inversion symmetry of the solutions. Examination of eq. (3.24) shows that if $x(t)$ is the response to a road profile $y(t)$, then $-x(t)$ is the response to the profile $-y(t)$. Also, eq. (3.34) reveals that, for uniformly bounded substrates, if $y(t)$ is an allowed substrate, then $-y(t)$ is also an allowed substrate. Let T_0 be the greatest continuous duration for which the acceleration exceeds zero, as the vehicle moves along any road profile from a given excitation set $Y(\hat{y})$ of uniformly bounded road profiles. From the symmetry of the differential equations describing the oscillation of the vehicle it is recognized that T_0 is also the greatest duration for which the acceleration is negative. Consequently the acceleration must change sign at least once in T_0 seconds. In other words, $1/T_0$ is the greatest lower bound for the number of acceleration sign changes per second, during motion on a uniformly bounded substrate.

We wish to determine the greatest continuous duration throughout which the vertical acceleration of the center of mass exceeds a given threshold value, given the initial conditions of eq. (3.25). We will approach this problem by investigating whether there exists a profile for which the vertical acceleration exceeds the threshold during a given interval. By varying the interval of interest we will determine the greatest continuous duration for which the acceleration exceeds the threshold.

The acceleration history during a time interval can be expressed in terms of the road profile during that interval and the state of the system at the onset of the interval. Our task is to determine if there exists a uniformly bounded road profile in the interval $[t_1, t_2]$ such that

$$x(t_1) = x_0 \qquad |y(t)| \leq \hat{y} \qquad y(t_1) = y(t_2) = y'(t_1) = y'(t_2) = 0 \tag{3.38}$$

$$a(t_1) = a(t_2) = a_0 \qquad a(t) \geq a_0 \qquad t_1 < t < t_2$$

By varying t_1 and t_2 we can find the greatest duration for which the acceleration exceeds a_0, starting from a given initial displacement x_0. The state (x_0, a_0) of the system at t_1 must be consistent with the initial conditions at time $t = 0$ specified in eq. (3.25). At the end of this section we will demonstrate a simple method for determining the set of all values (x_0, a_0) which can be reached at any time t_1.

A comment should be made on the end-point constraints on the acceleration: $a(t_1) = a(t_2) = a_0$. If a road profile exists which satisfies $a(t_1) > a_0$ or $a(t_2) > a_0$ as well as the other constraints in (3.38), then $[t_1, t_2]$ is certainly not the *longest* interval for which such a profile exists. Consequently, since we are seeking the greatest *continuous* duration for which $a(t) \geq a_0$, it is sufficient to require equality constraints rather than inequality constraints at the end-points.

We can formulate this as an optimization problem. Let $L(y, t)$ be a continuous functional of y whose form will be chosen later, and let

$$J = \int_{t_1}^{t_2} L(y, t)\, dt \tag{3.39}$$

Let S be the set of all road profiles for which the road profiles and the response satisfy the constraints in (3.38). S may be empty or may contain one or more elements. In any case we can ask for the supremum of J on the set S. No such supremum exists if S is empty. $L(y,t)$ is, by definition, a continuous function of y. Also, $Y(\hat{y})$ is a compact set. A theorem of functional analysis states that a continuous function on a compact set achieves a maximum value on that set (Kelly and Weiss, 1979). Hence, if S is not empty, then a supremum exists because S is a subset of all uniformly bounded profiles, $Y(\hat{y})$, and $L(y,t)$ is continuous on the compact set $Y(\hat{y})$. To re-iterate, if a road profile exists which satisfies the constraints in (3.38), then a road profile exists which satisfies (3.38) and which maximizes J subject to the constraints in (3.38).

Now, the maximization of J subject to (3.38) is an optimization problem whose solution can be sought using the technique of Euler-Lagrange. If a solution of the Euler-Lagrange equations exists, then a road profile satisfying (3.38) exists; if no solution of the Euler-Lagrange equations exists, then no road profile satisfying (3.38) exists. In the former case the greatest duration for which the acceleration exceeds a_0 is at least as great as $t_2 - t_1$. In the latter case $t_2 - t_1$ exceeds this greatest duration.

This optimization problem is complicated by the constraints being inequalities. Introduction of additional unknown functions, μ, η and ξ, converts the constraints to equalities (Vanderplaats, 1984). Consider the maximization of

$$J = \int_{t_1}^{t_2} L(y, \mu, \eta, \xi, t) dt \tag{3.40}$$

subject to the constraints:

$$f_1(y, \mu, t) \equiv y(t) - \mu^2 + \hat{y} = 0 \tag{3.41}$$

$$f_2(y, \eta, t) \equiv y(t) + \eta^2 - \hat{y} = 0 \tag{3.42}$$

$$f_3(y, \xi, t) \equiv a_y(t) - \xi^2 - a_0 = 0 \tag{3.43}$$

We must find four real functions, $y(t), \mu(t), \eta(t)$ and $\xi(t)$, to maximize J subject to eqs. (3.41) – (3.43). To solve this we employ the calculus of variations (Weinstock, 1974). The constraints are adjoined to the integrand of eq. (3.40) with the introduction of three unspecified multiplier functions $\lambda_1(t), \lambda_2(t)$ and $\lambda_3(t)$. The seven unknown functions must satisfy the three constraint equations and the four Euler-Lagrange equations during the interval $[0, T]$, to be formulated as follows.

The vertical acceleration of the center of mass, on the road profile $y(t) = y_1(t) = y_2(t)$, given $x(t_1) = x_0, a(t_1) = a_0$, $y(t_1) = y'(t_1) = 0$ is written as

$$\begin{aligned} a_y(t) &= \omega_o^2 e^{-\zeta\omega_o t}\left[\left(\frac{\zeta^2}{\beta} - \beta\right)\left(x_0\zeta + \frac{v_0}{\omega_o}\right)\sin\beta\omega_o t\right. \\ &+ \left.\left(x_0(\beta^2 - \zeta^2) - \frac{2\zeta v_0}{\omega_o}\right)\cos\beta\omega_o t\right] + 2\zeta\omega_o y'(t) \\ &+ \omega_o^2\left(1 - 4\zeta^2\right)y(t) + 2\int_0^t y(\tau)\rho(t-\tau)d\tau \end{aligned} \tag{3.44}$$

where the initial velocity, v_0, is related to the other initial conditions, x_0 and a_0, by

$$v_0 = \frac{x_0\omega_o}{2\zeta}\left(1 - 2\zeta^2\right) - \frac{a_0}{2\zeta\omega_o} \tag{3.45}$$

Define

$$F(y, y', t) = L + \sum_{m=1}^{3} \lambda_m f_m \tag{3.46}$$

Consider the cost function

$$L = \mu^2(t) + \eta^2(t) + \xi^2(t) - y^2(t) \tag{3.47}$$

The Euler-Lagrange equations are

$$0 = \frac{\partial F}{\partial y} - \frac{d}{dt}\left(\frac{\partial F}{\partial y'}\right) = -2y(t) + \lambda_1(t) + \lambda_2(t) + 2\lambda_3(t)\psi(t) \tag{3.48}$$

$$0 = \frac{\partial F}{\partial \mu} - \frac{d}{dt}\left(\frac{\partial F}{\partial \mu'}\right) = 2[1 - \lambda_1(t)]\mu(t) \tag{3.49}$$

$$0 = \frac{\partial F}{\partial \eta} - \frac{d}{dt}\left(\frac{\partial F}{\partial \eta'}\right) = 2[1 + \lambda_2(t)]\eta(t) \tag{3.50}$$

$$0 = \frac{\partial F}{\partial \xi} - \frac{d}{dt}\left(\frac{\partial F}{\partial \xi'}\right) = 2[1 - \lambda_3(t)]\xi(t) \tag{3.51}$$

The systematic solution of these equations is discussed in section 3.2.8.

We now consider the determination of the set of values (x, v) which can be reached at time t by the system in eq. (3.24) starting with the conditions in eq. (3.25). Eq. (3.45) relates this set to the set of values (x, a). Let $R(t)$ be the set of displacement and velocity values $(x(t), v(t))$ which can be reached at time t, starting from the initial conditions $x(0) = v(0) = 0$ and traveling over uniformly bounded road profiles $|y(t)| \leq \hat{y}$. That is

$$R(t) = \left\{ r : r(t) = \begin{pmatrix} x_y(t) \\ v_y(t) \end{pmatrix} \quad \text{for all} \quad y \in Y(\hat{y}) \right\} \tag{3.52}$$

where x_y and v_y are given by eqs. (3.28) and (3.30) respectively, with $y = y_1 = y_2$. $R(t)$ is a convex set, so to plot $R(t)$ we need only determine its boundary points. $R(t)$ contains the origin, so the boundary of $R(t)$ is the locus of elements of $R(t)$ which are farthest along rays from the origin. Let $u(\theta) = (\cos\theta, \sin\theta)$ be a unit vector in the (x, v) plane at the angle θ from the x axis. The line perpendicular to $u(\theta)$ at a distance D from the origin consists of points (x, v) satisfying $x\cos\theta + v\sin\theta = D$. Thus the distance from the origin of a line perpendicular to $u(\theta)$ and tangent to $R(t)$ is

$$D(\theta) = \max_{y \in Y(\hat{y})} \{x_y(t)\cos\theta + v_y(t)\sin\theta\} \tag{3.53}$$

$$= 2\hat{y}\zeta\omega_o + 2\max_{y \in Y(\hat{y})} \int_0^t y(\tau)\{\varphi(t-\tau)\cos\theta + \psi(t-\tau)\sin\theta\}\, d\tau \tag{3.54}$$

The maximum of the integral in eq. (3.54) is obtained by choosing $y(\tau)$ equal in magnitude to $\hat{y}$ and equal in sign to $\varphi(t-\tau)\cos\theta + \psi(t-\tau)\sin\theta$. Let $y_\theta(\tau)$ denote this road profile. That is,

$$y_\theta(\tau) = \hat{y}\ \mathrm{sgn}[\varphi(t-\tau)\cos\theta + \psi(t-\tau)\sin\theta] \tag{3.55}$$

The line perpendicular to $u(\theta)$ and tangent to $R(t)$ contacts $R(t)$ at the point $(x_{y_\theta}, v_{y_\theta})$. Hence this point is a boundary point of $R(t)$. Employing eqs. (3.28) and (3.30), the boundary points (x, v) of $R(t)$ are found by varying θ on $[0, 2\pi]$ in the relations

$$x = 2\hat{y} \int_0^t \operatorname{sgn}[\varphi(t-\tau)\cos\theta + \psi(t-\tau)\sin\theta]\varphi(\tau)d\tau \tag{3.56}$$

$$\begin{aligned} y &= 2\zeta\omega_o\hat{y}\operatorname{sgn}[\varphi(0)\cos\theta + \psi(0)\sin\theta] \\ &\quad + 2\hat{y}\int_0^t \operatorname{sgn}[\varphi(t-\tau)\cos\theta + \psi(t-\tau)\sin\theta]\psi(\tau)d\tau \end{aligned} \tag{3.57}$$

To recapitulate, we state in algorithmic terms the results of this section. The greatest duration for which the acceleration exceeds a given threshold value, a_0, is evaluated by the following numerical procedure:

1. Choose the time of initiation, t_1. Clearly t_1 must satisfy $\hat{a}_{ub}(t_1) > a_0$.

2. Choose the end-point for the time interval: $t_2 = t_1 + n\,\delta T$, $n = 1, 2, \ldots$, where δT is a time increment.

3. Choose a value of (x_0, v_0) in $R(t_1)$. If $R(t_1)$ has been thoroughly sampled for the current value of t_2, go to step 7.

4. Seek a solution of the Euler-Lagrange equations, (3.41) - (3.43) and (3.48) - (3.51).

5. If a solution in step 4 exists, return to step 2 and increase t_2 by δT.

6. If no solution in step 4 exists, return to step 3 and select a new element of $R(t_1)$.

7. Adopt the previous value of $t_2 - t_1$ as the greatest duration of acceleration at the current time of initiation, t_1. Denote this value $T(t_1)$.

8. Return to step 1 and replace t_1 by $t_1 + \delta t$ if $R(t_1)$ has not converged to a constant set. If $R(t_1)$ has converged, then the greatest duration of acceleration is the maximum of the sequence of local maxima obtained: $T(t_1), T(t_1 + \delta t), T(t_1 + 2\delta t), \ldots$.

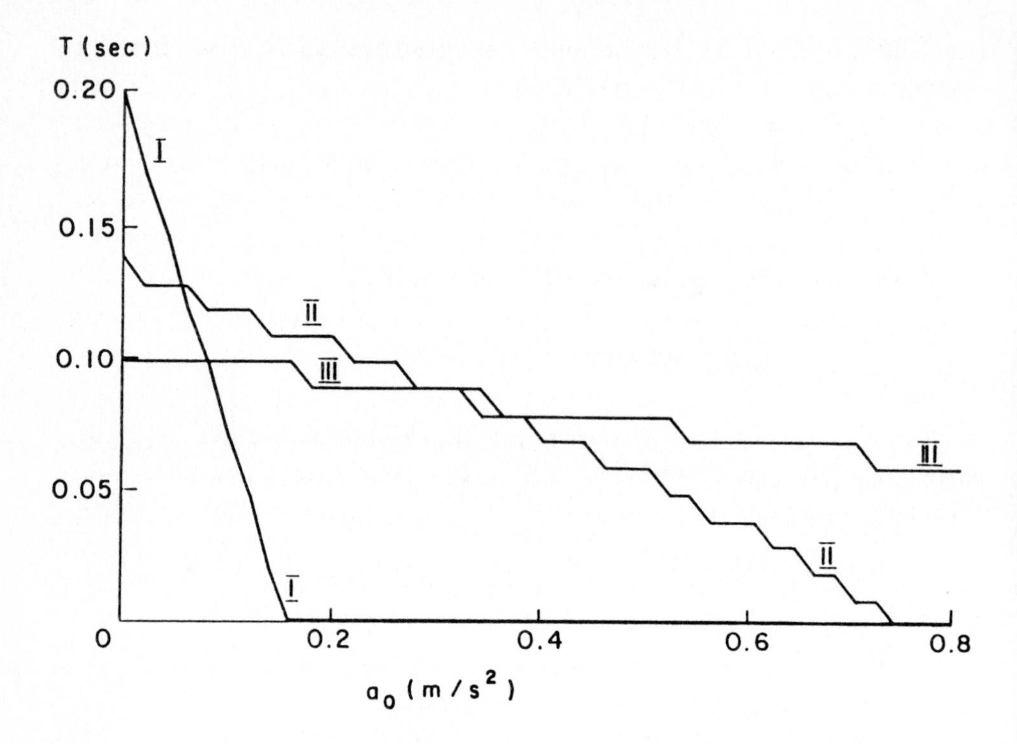

Figure 3.7: Maximum duration of acceleration excursion from $t = 0$ versus the acceleration threshold, for three values of the stiffness coefficient.

A simpler version of this algorithm occurs if one wishes to determine the greatest continuous duration of acceleration above a given threshold, a_0, starting at time $t = 0$. In this case the loop on t_1 is avoided and the set $R(t_1)$ contains only a single element: The initial conditions of the vehicle motion.

Fig. 3.7 shows the greatest continuous duration of acceleration excursion versus the acceleration threshold, starting at $t = 0$ from the initial conditions in eq. (3.25) and for $\hat{y} = 0.01$m. Results are shown for three different values of the stiffness coefficient k. It is seen that the duration of excursion is very sensitive to the stiffness of the suspension system. In curve II the stiffness coefficient is 4×10^4 N/m, which is the value used in Newland's work (1986) and elsewhere in this section. This curve shows, for example, that 0.11s is the greatest continuous duration for which the acceleration can be maintained above 0.2m/s^2, starting at time $t = 0$ from the initial conditions in eq. (3.25). Likewise, the same curve indicates that 0.042s is the greatest possible duration of acceleration above 0.6m/s^2. Curve II reaches the axis at $a_0 = 0.76\text{m/s}^2$, indicating that accelerations greater than this value can not be maintained for a non-zero duration starting from time $t = 0$. This agrees with eq. (3.37) and Fig. 3.4 which indicate that the upper bound of the acceleration at $t = 0$ is $\hat{a}_{ub} = 2\hat{y}\omega_0^2 \left|\frac{1}{2} - 2\zeta^2\right| = 0.76\text{m/s}^2$ for $\hat{y} = 0.01\text{m}$.

When the stiffness is increased to 8×10^4 N/m (curve III) the greatest duration of acceleration-excursion above 0.6m/s^2 is 0.071s, nearly double the value in curve II. This is due to the fact that the stiffer suspension system is able to attain great accelerations. Conversely, comparison of curves II and III indicates that the stiff suspension system maintains low values of the acceleration for shorter durations than for the suspension of moderate stiffness.

Finally, the results in curve I indicate that a very soft suspension ($k = 1 \times 10^4$ N/m) achieves only comparatively low accelerations, but can maintain these acceleration-excursions for longer durations.

As mentioned at the beginning of this section, vertical vibrational acceleration resulting from motion on an irregular surface involves roughly periodic changes in the sign of the acceleration. Examination of Fig. 3.7 for duration of acceleration excursions above $a = 0$ indicates that the stiffest spring (curve III) tends to oscillate twice as rapidly as the softest spring (curve I). Furthermore, the frequencies involved (for the specific situation examined) are in the range of 5 to 10 sign-changes per second. Considerations such as these enable the designer to relate the vehicle performance on an irregular substrate to accepted limits for vibration exposure.

The step-like form of all three curves is due to the fact that "bang-bang" profiles produce excursions of maximal duration. As the acceleration threshold is reduced, the maximum duration of excursion remains constant until the changing boundary conditions allow an alteration in the "bang-bang" profile.

3.2.6 Substrate Profiles With Bounded Slopes

The previous three sections have been devoted to the study of vehicle dynamics on uniformly bounded substrates. In this section we introduce a different model for the substrate profile, by assuming that the slope rather than the height of the substrate is uniformly bounded. This is a simple but instructive representation of rolling terrain and will be used in section 3.2.7 in studying vehicle response to barriers. Analytical expressions are obtained for the envelopes of extremal responses of the vehicle.

The convex model of the substrate asserts that the absolute magnitude of the slope of the substrate can deviate from zero by no more than a specified amount. Furthermore, the two wheels of the trailer proceed over identical substrate profiles. It is also assumed, as before, that the forward velocity of the vehicle is constant and that the initial displacement and initial vertical velocity of the center of mass are zero. Consequently, the vertical displacements imposed on the vehicle, $y_1(t)$ and $y_2(t)$ in eq. (3.24), each belong to the following set:

$$S(\hat{s}) = \left\{ y : \ y(0) = 0 \quad , \quad \left| \frac{dy}{dt} \right| \leq \hat{s} \quad , \quad 0 \leq t < \infty \right\} \tag{3.58}$$

Equations (3.28), (3.30) and (3.31) express the vehicle response. However, it should be noted that $y_1(t) = y_2(t)$ in this slope-bounded model, for otherwise the vertical displacement between the two wheels of the trailer could increase without bound. Also, it is convenient to express the vehicle responses in terms of the derivatives of the road profile, as follows:

$$x(t) = 2\int_0^t y'(\tau)\left[\Phi(t,t) - \Phi(\tau,t)\right] d\tau \tag{3.59}$$

$$v(t) = 2\int_0^t y'(\tau)\varphi(t-\tau)d\tau \tag{3.60}$$

$$a(t) = 2\omega_o\zeta y'(t) + 2\int_0^t y'(\tau)\psi(t-\tau)d\tau \tag{3.61}$$

where y' is the time derivative of y, φ and ψ are defined in eqs. (3.29) and (3.32) and $\Phi(\tau, t)$ is defined as

$$\Phi(\tau, t) = \int_0^{\tau} \varphi(t - z) dz \tag{3.62}$$

The extremal responses are obtained directly from these expressions by an argument similar to that employed in deriving eqs. (3.35) to (3.37). Examination of eq. (3.59) reveals that the extremal displacement $\hat{x}_{sb}(t)$ at time t due to slope bounded terrain is obtained from the road whose slope is always of maximal magnitude, and of sign equal to the sign of $\Phi(t,t) - \Phi(\tau, t)$. Thus the envelope of extremal displacements is

$$\hat{x}_{sb}(t) = 2\hat{s} \int_0^t |\Phi(t,t) - \Phi(\tau, t)| \, d\tau \tag{3.63}$$

Similar arguments applied to eqs. (3.60) and (3.61) yield expressions for the greatest possible vertical velocity and acceleration at time t as

$$\hat{v}_{sb}(t) = 2\hat{s} \int_0^t |\varphi(\tau)| d\tau \tag{3.64}$$

$$\hat{a}_{sb}(t) = 2\hat{s}\zeta\omega_o + 2\hat{s} \int_0^t |\psi(\tau)| d\tau \tag{3.65}$$

A remarkable conclusion can be drawn by comparing these results with eqs. (3.35) and (3.36): It turns out that the extremal responses on slope-bounded and uniformly-bounded substrates are related as follows:

$$\hat{v}_{sb}(t) = \frac{\hat{s}}{\hat{y}} \hat{x}_{ub}(t) \tag{3.66}$$

$$\hat{a}_{sb}(t) = \frac{\hat{s}}{\hat{y}} \hat{v}_{ub}(t) \tag{3.67}$$

In other words, the maximum velocity of a vehicle on a slope-bounded profile is proportional to the maximum displacement of a vehicle on a uniformly bounded profile. Similarly, the maximum acceleration of the vehicle on the slope-bounded terrain is proportional to the maximum velocity on the uniformly bounded profile.

3.2.7 Isochronous Obstacles

The results of section 3.2.6 can be readily adapted to the study of a simple model for vehicle behaviour on a barrier. The aim of our analysis is, as before, to derive expressions for the extremal responses as the vehicle passes over any barrier from among an ensemble of barriers. The maximum vibrational acceleration is related to the peak acceleration allowed for preservation of safety, work-efficiency or comfort.

In practice, the time which a vehicle expends in traversing a barrier depends in a complicated way on the shape and traction-properties of the barrier and on the characteristics of the engine in response to a varying load. However, as a simple approximation to these phenomena, we will assume that the duration which the vehicle spends on the obstacle is the same for all allowed obstacles. Furthermore, in evaluating the vertical response of the center of mass we will assume that both wheels of the trailer follow the same obstacle profile, so $y_1(t) = y_2(t)$ in eq. (3.24). Finally, it appears logical to define the obstacles in terms of envelope-bounds on their slopes. This means that the obstacle profile-functions belong to the following set:

$$B(s_1(t), s_2(t)) = \left\{ y : \ y(0) = 0 \ \ ; \ \ s_1(t) \le \frac{dy}{dt} \le s_2(t) \right\} \tag{3.68}$$

where $s_1(t)$ and $s_2(t)$ are specified functions for which $s_1(t) \le s_2(t)$ during $0 \le t \le D$, and $s_1(t) = s_2(t) = 0$ for $t > D$. D is the duration of passage over the obstacle. The barriers in this ensemble are isochronous in the sense that the duration of passage is the same for all members of the set B.

Fig. 3.8 shows typical slope-envelope functions for unimodal isochronous barriers. All the barriers in the ensemble exhibit positive slopes from $t = 0$ to $t = 2.5$ sec and negative slopes from $t = 2.5$ to $t = 5$ sec. Thus all the barriers in the ensemble display a single peak which is reached at $t = 2.5$ sec. We have chosen the functions $s_i(t)$ for this case as

$$s_1(t) = a_1 \sin \frac{2\pi t}{5} \qquad 0 \le t \le 2.5 \tag{3.69}$$

$$= a_2 \sin \frac{2\pi t}{5} \qquad 2.5 \le t \le 5 \tag{3.70}$$

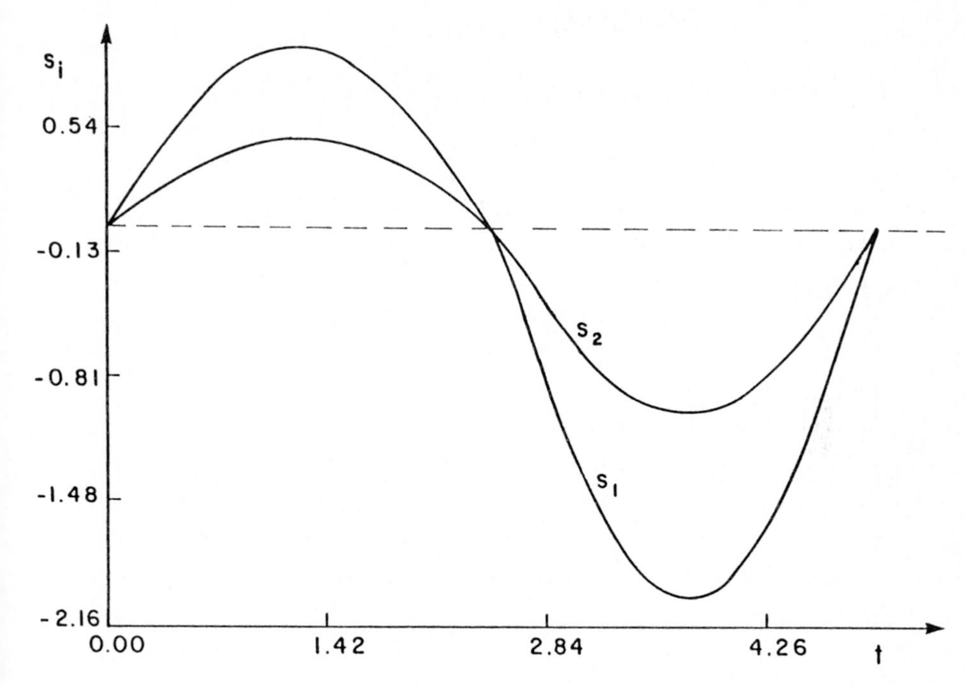

Figure 3.8: Envelope functions for a typical unimodal isochronous obstacle.

$$s_2(t) = \sin\frac{2\pi t}{5} \quad 0 \leq t \leq 5 \tag{3.71}$$

The slope-envelope functions shown in Fig. 3.9 illustrate that it is not necessary for the barriers to have a fixed number of peaks. The barriers in this ensemble display positive slope during the interval [0, 2] and negative slope during the interval [3, 5]. The intermediate time interval, [2, 3], corresponds to the top region of the barrier, which can be choppy or wavey. Multi-modal isochronous barriers are modelled here with slope-envelopes:

$$s_1(t) = \frac{t}{3}(2-t)(5-t) \tag{3.72}$$

$$s_2(t) = \frac{t}{3}(3-t)(5-t) \tag{3.73}$$

Fig. 3.10 illustrates, as an example from among the infinity of possibilities, the slope profile of a wavey-topped barrier whose slope is

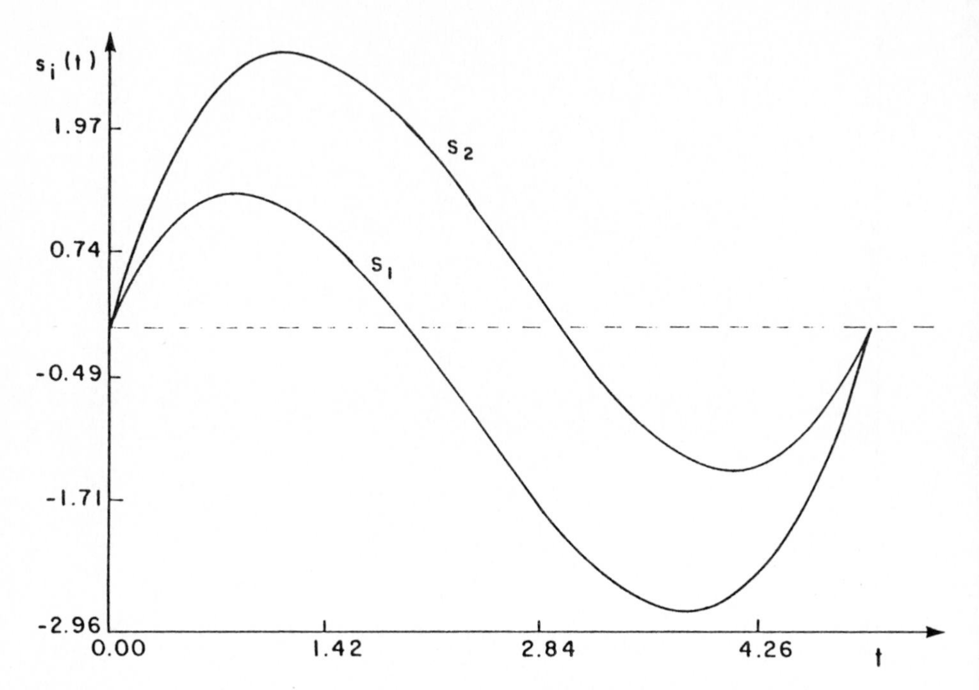

Figure 3.9: Envelope functions for a typical multi-modal isochronous obstacle.

bounded by $s_1(t)$ (eq. 3.72)) and $s_2(t)$ (eq. 3.73)). The slope of this barrier is given by

$$s(t) = s_1(t)\sin^2 2\pi t + s_2(t)\cos^2 2\pi t$$

The corresponding height profile is shown in Fig. 3.11.

The vertical displacement, velocity and acceleration of the center of mass of the vehicle are described by eqs. (3.59) to (3.61). Extremal values of these functions are obtained by defining the instants at which the functions $\Phi(t,t) - \Phi(\tau,t)$, $\varphi(\tau)$ and $\psi(\tau)$ change sign. Specifically, consider the following sets of instants:

$$X_+(t) = \{\tau :\ 0 \le \tau \le t \quad , \quad \Phi(t,t) - \Phi(\tau,t) \ge 0\} \tag{3.74}$$

$$X_-(t) = \{\tau :\ 0 \le \tau \le t \quad , \quad \Phi(t,t) - \Phi(\tau,t) < 0\} \tag{3.75}$$

Thus $X_+(t)$ is the set of time-instants during $[0,t]$ at which $\Phi(t,t) - \Phi(\tau,t)$ is non-negative, while $X_-(t)$ is the set of instants at which

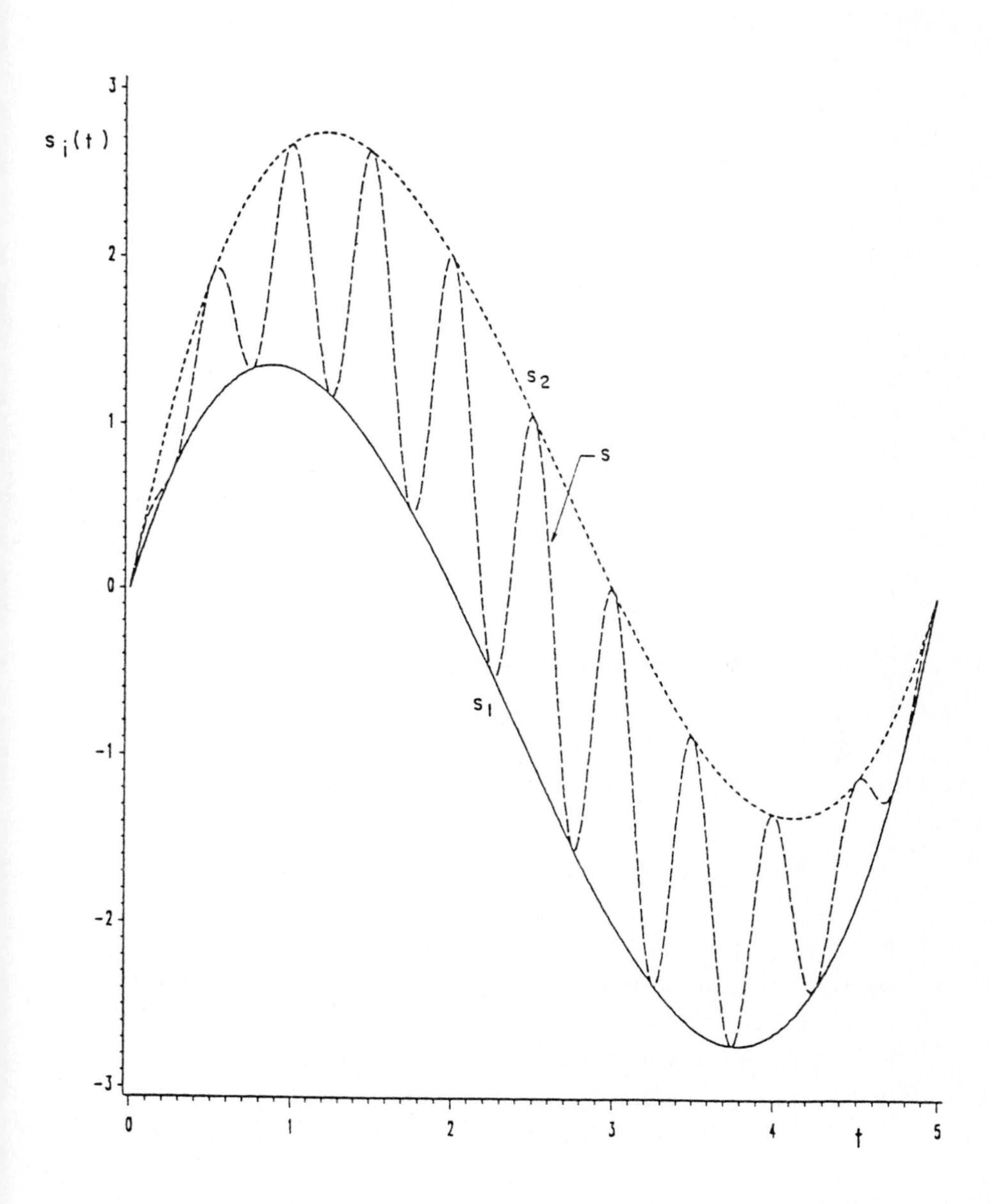

Figure 3.10: Slope profile of a wavey-topped isochronous barrier.

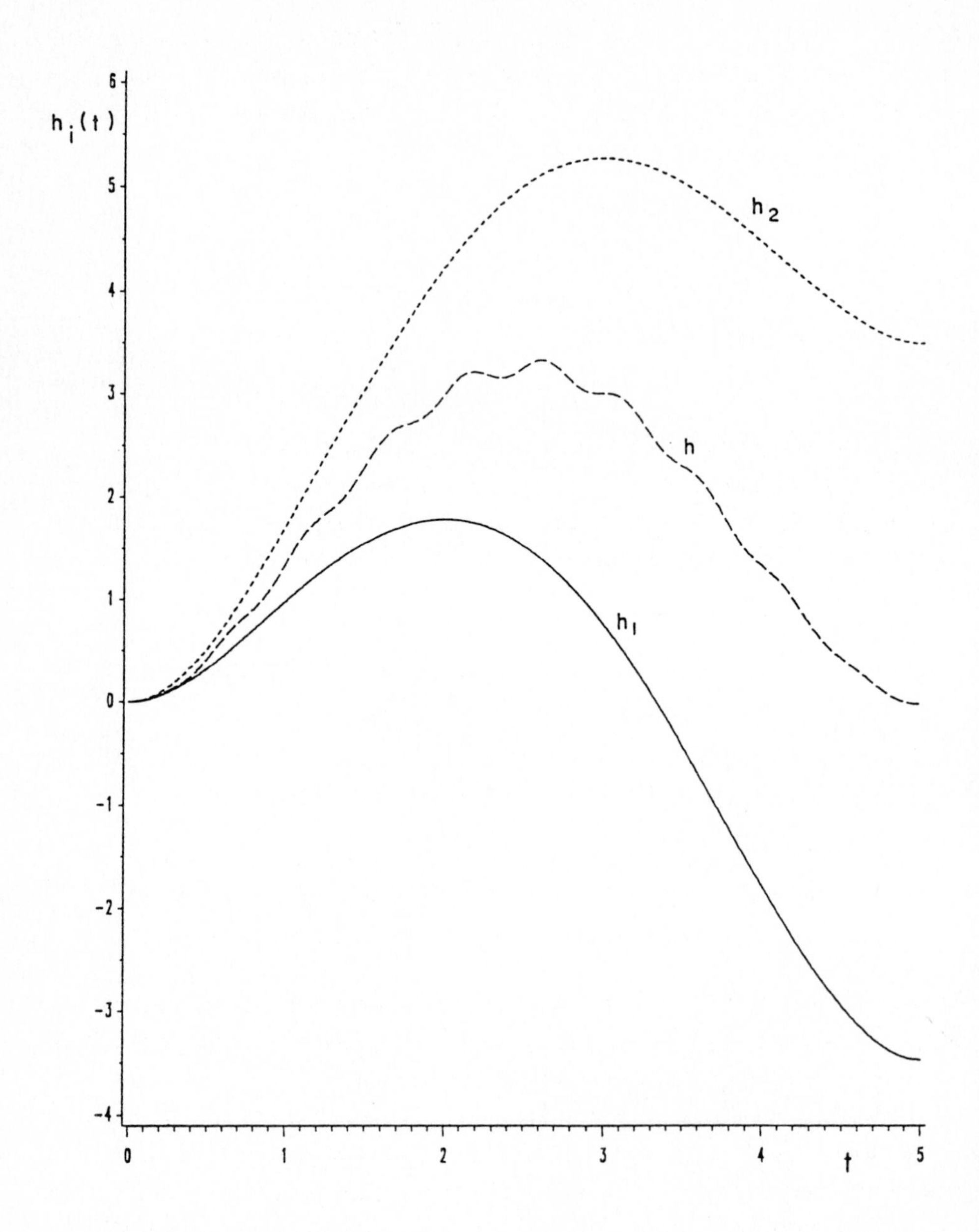

Figure 3.11: Height profile of a wavey-topped isochronous barrier.

$\Phi(t,t) - \Phi(\tau,t)$ is negative. Similarly, $V_+(t)$ and $V_-(t)$ are sets containing the instants in $[0,t]$ at which $\varphi(t-\tau)$ is non-negative and negative, respectively, while $A_+(t)$ and $A_-(t)$ contain the instants at which $\psi(t-\tau)$ is non-negative and negative:

$$V_+(t) = \{\tau : \ 0 \le \tau \le t \quad , \quad \varphi(t-\tau) \ge 0\} \tag{3.76}$$

$$V_-(t) = \{\tau : \ 0 \le \tau \le t \quad , \quad \varphi(t-\tau) < 0\} \tag{3.77}$$

$$A_+(t) = \{\tau : \ 0 \le \tau \le t \quad , \quad \psi(t-\tau) \ge 0\} \tag{3.78}$$

$$A_-(t) = \{\tau : \ 0 \le \tau \le t \quad , \quad \psi(t-\tau) < 0\} \tag{3.79}$$

The maximum responses are obtained from eqs. (3.59) to (3.61) by choosing the barrier profile whose slope at each instant is either s_1 or s_2, depending on the sign of the rest of the integrand. Explicitly, the maximum vertical displacement, velocity and acceleration of the center of mass at time $t \ge 0$ are

$$\hat{x}_{io}(t) = 2\int_{X_+(t)} s_2(\tau)[\Phi(t,t)-\Phi(\tau,t)]d\tau + 2\int_{X_-(t)} s_1(\tau)[\Phi(t,t)-\Phi(\tau,t)]d\tau \tag{3.80}$$

$$\hat{v}_{io}(t) = 2\int_{V_+(t)} s_2(\tau)\varphi(t-\tau)d\tau + 2\int_{V_-(t)} s_1(\tau)\varphi(t-\tau)d\tau \tag{3.81}$$

$$\hat{a}_{io}(t) = 2\omega_o\zeta s_2(t) + 2\int_{A_+(t)} s_2(\tau)\psi(t-\tau)d\tau + 2\int_{A_-(t)} s_1(\tau)\psi(t-\tau)d\tau \tag{3.82}$$

It should be noted that these expressions describe the maximum possible responses at any time t during and after passing the obstacle. The subscript "io" represents "*i*sochronous *o*bstacle."

Fig. 3.12 shows the vertical acceleration of the center of mass in passage over the unimodal isochronous barrier whose envelope functions are shown in Fig. 3.8. Various values are shown for the coefficients a_1 and a_2. Similarly, Fig. 3.13 shows the vertical acceleration of the center of mass in passage over a multi-modal isochronous barrier whose envelope functions are shown in Fig. 3.9.

In this section convex sets of functions have been used to describe uncertainty in the profile of irregular terrain. Two classes of convex

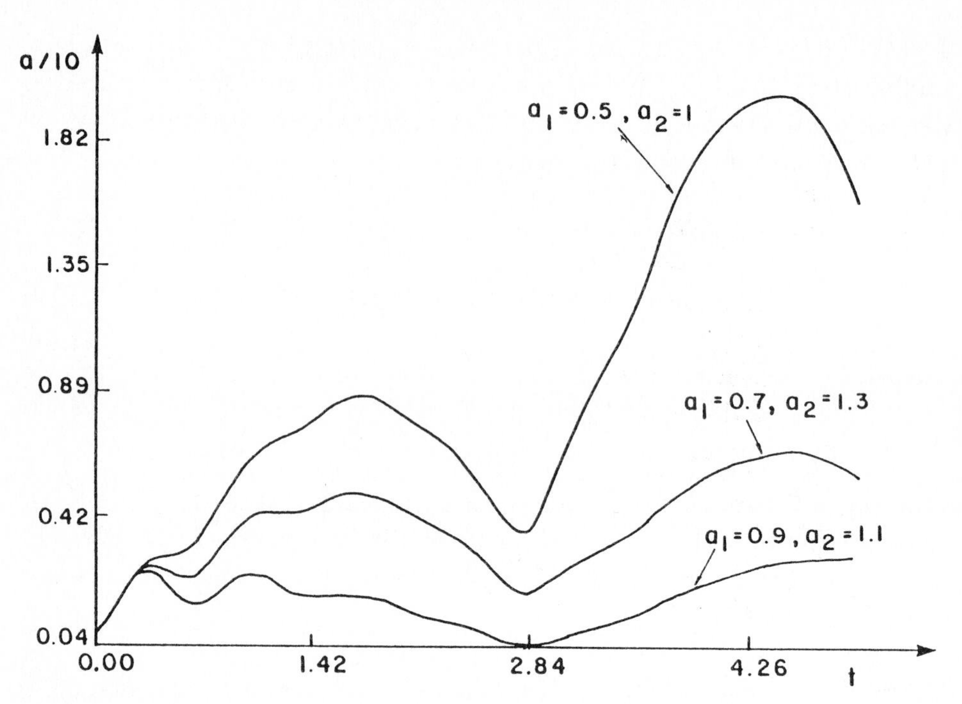

Figure 3.12: Vertical acceleration for passage over a unimodal isochronous obstacle.

models have been employed. Uniform bounds on the terrain profile represent irregular surfaces with moderate roughness. Slope bounds represent rolling terrain and isolated obstacles. The vibrational acceleration induced in a vehicle passing over such surfaces is an important performance parameter which must be considered in vehicle analysis and design. The vertical acceleration of a two-wheeled trailer has been analyzed in a number of situations, with a linear model of the vibration dynamics. The following conclusions can be drawn.

1. The maximum vertical acceleration, for any terrain in a uniformly bounded set of surface profiles, can be readily evaluated. The peak vertical acceleration for the vehicle analyzed here can be as large as 5.3m/s^2 on a substrate whose height varies by ± 0.01m. (See Fig. 3.4). Accepted standards (Harris, 1987) indicate that this results in appreciable discomfort and substantial reduction of work efficiency in a matter of minutes.

2. A methodology has been developed for evaluating the maxi-

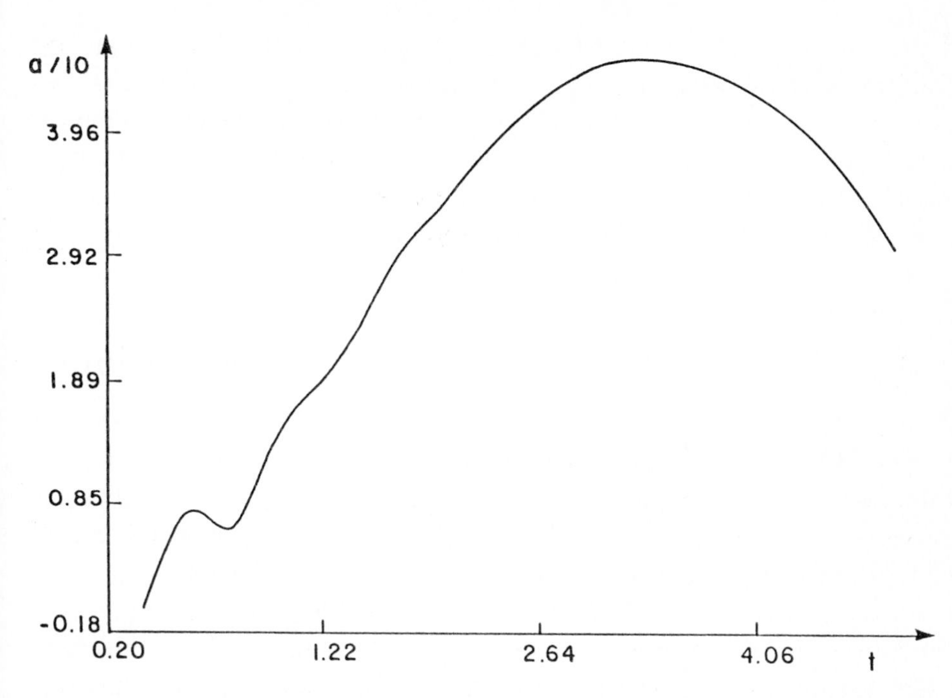

Figure 3.13: Vertical acceleration for passage over a multi-modal isochronous obstacle.

mum duration for which the vibrational acceleration exceeds a given threshold. This is useful in relating the complex oscillatory motion on an irregular and imperfectly known surface to the accepted limits of vibration exposure which are based on single-frequency excitation. Furthermore, $1/T_0$ is the greatest lower bound of the number of acceleration sign-changes per second, if T_0 is the greatest duration for which the acceleration exceeds zero. For the two-wheeled trailer on a uniformly-bounded substrate it is shown quantitatively the extent to which the spring constant influences the maximum duration of acceleration excursions. For spring constants from 1×10^4 to 8×10^4 N/m the maximum duration of positive acceleration varies from 0.2 to 0.1 sec for the situation examined. This corresponds to variation in the minimum number of acceleration sign-changes per second from 5 s^{-1} (for the soft spring) to 10 s^{-1} (for the hard spring). (See Fig. 3.7).

3. In many field situations the terrain profile includes transient substrate irregularities. The dynamics of vehicle response is quite different in passing an isolated obstacle than on moderately rough but uniformly bounded terrain. It is shown that uncertainty in barrier profiles can be conveniently represented with the set-theoretical technique of convex modelling. It is shown that the maximum acceleration is readily evaluated in passing over a range of unimodal and multi-modal barriers. Quite large accelerations were found for most of the numerical cases examined. (See Figs. 3.12 and 3.13).

3.2.8 Solution of the Euler-Lagrange Equations

In this section we discuss the method by which eqs. (3.41) – (3.43) and (3.48) – (3.51) are solved to determine the existence of a uniformly bounded road profile during a fixed time interval $[0, T]$, such that the vertical acceleration exceeds a threshold value. The maximum duration of an acceleration excursion is found by employing the algorithm outlined in section 3.2.5. The solution develops differently in time under the influence of the various acceleration and road-profile constraints. Four distinct cases must be considered. Furthermore, the non-linearity of the equations and of the constraints allows for a finite number of possible solutions. The branch-points of the solution occur as the solution reaches or moves away from a boundary point of the solution region. The numerical solution of these equations follows all possible solutions.

Case 1: a_y and y at interior points: $a_y > a_0$ and $|y| < \hat{y}$. From the constraint eqs. (3.41) – (3.43) one learns that the functions μ, η and ξ all differ from zero if and only if y and a_y are not at their boundary values. In this case the multipliers λ_n are obtained from eqs. (3.49) – (3.51) as $\lambda_1 = -\lambda_2 = \lambda_3 = 1$. Thus eq. (3.48) implies that $y(t)$ must satisfy

$$y(t) = \psi(t) \tag{3.83}$$

provided that

$$|y| < \hat{y} \quad \text{and} \quad a_y > a_0 \tag{3.84}$$

Eqs. (3.44) and (3.83) establish the values of $y(t)$ and $a_y(t)$ until a boundary value is reached by either y or a_y.

Case 2: y at boundary; a_y at interior point: $a_y > a_0$ and $|y| = \hat{y}$. If y reaches its boundary then either (3.49) or (3.50) (depending on which boundary is reached) ceases to determine the value of λ_1 or λ_2. However, since y is now known, the undetermined λ is determined by eq. (3.48). The value of the road profile remains fixed at the boundary value until the righthand side of eq. (3.83) returns to within the range $\pm\hat{y}$. Now the equations can be solved by either of two alternative developments of the profile y : The profile can either continue along the boundary or leave the boundary by following eq. (3.83). Both solutions are allowed by the Euler-Lagrange equations. Since y must be continuous, it can move into the interior only at those instants at which the righthand side of eq. (3.83) crosses the boundary.

Case 3: a_y at boundary; y at interior point: $a_y = a_0$ and $|y| < \hat{y}$. If a_y reaches its boundary while y is still in its interior region, then λ_3 is no longer determined by eq. (3.51) and the further development of y is determined by eq. (3.44), with a_y fixed at a_0. Again the equations can be satisfied by either of two alternative developments of the profile. One solution keeps a_y at a_0 and has y determined by eq. (3.44). The other solution is that y returns to the form of eq. (3.83), which is possible if the resulting acceleration exceeds a_0 and if the road profile is continuous at the transition between the solutions of eqs. (3.44) and (3.83).

Case 4: a_y and y both at boundary: $a_y = a_0$ and $|y| = \hat{y}$. If a_y and y both assume boundary values, then the solution can continue only if either a_y or y or both move away from their boundaries, since eq. (3.44) will not allow both a_y and y to be constant. Thus the solution returns to cases $1 - 3$.

The numerical development of the solution from $t = 0$ terminates if the endpoint time, T, is reached, indicating that the acceleration can be maintained above the threshold value for the duration T. Alternatively, if no solution is available at some intermediate instant, then the numerical development of the solution terminates, indicating that the acceleration can not be maintained above the threshold value for the duration T.

3.3 Seismic Excitation

The problem of the earthquake excitation of structures has, deservingly, attracted many investigators. Numerous papers, monographs and specialized journals are available on this topic. Much of this effort has uti-

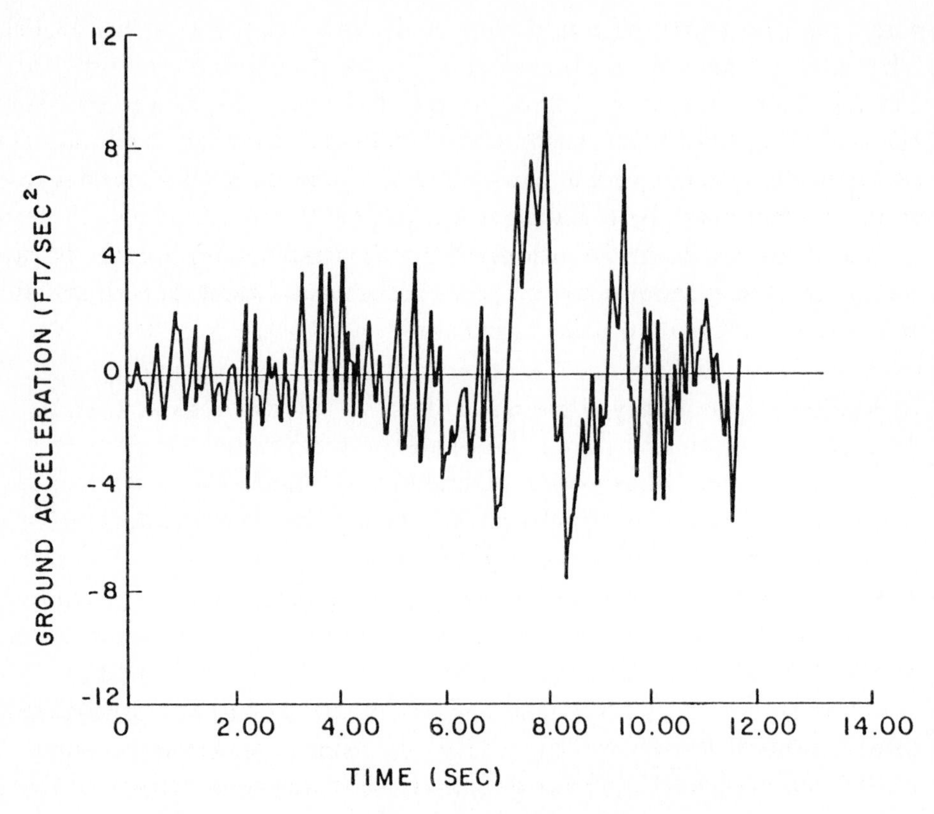

Figure 3.14: A typical earthquake accelerogram.

lized a probabilistic approach, modelling the earthquake excitation as a random process. Fig. 3.14 shows a typical earthquake accelerogram illustrating the highly complex temporal variation of the excitation. Another approach, the so-called "critical excitations" method, was pioneered by Drenick in 1968. Later on, a number of papers and reports extended this novel approach (see papers by Drenick, 1970, 1984; Shinozuka, 1970; and others). The main idea of these papers is to use highly reliable but limited deterministic information characterizing the ground motion. This information is written as a constraint:

$$F(x(t)) \leq M^2 \tag{3.85}$$

where $x(t)$ is some characteristic of the earthquake excitation, $F(\cdot)$ is a functional and M^2 is a positive constant. This is supplemented by a governing differential equation:

$$P(y(t)) = Q(x(t)) \tag{3.86}$$

where $P(\cdot)$ and $Q(\cdot)$ are differential operators. One then looks for an excitation function $x(t)$ which maximizes the response $y(t)$ on the set of allowable excitations, defined in eq. (3.85).

The question is how to formulate the constraint in eq. (3.85). Information on strong ground motion is rather scant. According to Vanmarcke (1976), earthquakes can be characterized by the following quantities:

1. Duration of the strong ground motion.
2. Ground motion intensity.
3. Envelope of the amplitude spectrum.
4. Effects of the macro-zone and the micro-zone.
5. Effect of focal distance.

Drenick, Novomestky and Bagchi (1984) suggest adding a sixth item, namely bounds on the peak ground acceleration and velocity.

Various constraints on the ground motion appear in the literature. In his pioneering work, Drenick (1968, 1970) used a constraint on the total energy which the earthquake is likely to develop at the location of the structure:

$$F(x(t)) = \int_{-\infty}^{\infty} x^2(t)\, dt \tag{3.87}$$

In order to find the least favorable response, consider a single degree-of-freedom system. The response of such a system is written as follows:

$$y(t) = \int_{-\infty}^{\infty} h(t-\tau)x(\tau)\, d\tau \tag{3.88}$$

where $h(t)$ is the impulse response function. In order to find the maximum response, we utilize the Cauchy-Schwarz inequality:

$$\left(\int_{-\infty}^{\infty} h(t-\tau)x(\tau)d\tau\right)^2 \leq \int_{-\infty}^{\infty} h^2(t-\tau)d\tau \int_{-\infty}^{\infty} x^2(\tau)d\tau \tag{3.89}$$

with equality when $x(\tau)$ is proportional to $h(t-\tau)$. Thus a maximum in eq. (3.88) is attained if $x(\tau)$ is proportional to $h(t-\tau)$:

$$x(\tau) = \alpha h(t-\tau) \tag{3.90}$$

The maximizing excitation satisfies:

$$\int_{-\infty}^{\infty} x^2(\tau)\,d\tau = M^2 \tag{3.91}$$

or

$$\alpha^2 \int_{-\infty}^{\infty} h^2(t-\tau)\,d\tau = M^2 \tag{3.92}$$

and consequently:

$$\alpha = \frac{M}{\sqrt{\int_{-\infty}^{\infty} h^2(t-\tau)d\tau}} \tag{3.93}$$

Hence the least favorable excitation equals:

$$x_{lf}(\tau) = \frac{M}{\sqrt{\int_{-\infty}^{\infty} h^2(t-\tau)d\tau}} h(t-\tau) \tag{3.94}$$

Note that the excitation $x(\tau)$ during the interval $[0,t]$ which maximizes the response at time t is proportional to $h(t-\tau)$. In other words, since $y(t)$ is the convolution of $h(t)$ and $x(t)$, the response is maximized at time t when $x(\tau)$ is proportional to $h(t-\tau)$.

For convenience let us denote:

$$\int_{-\infty}^{\infty} h^2(t-\tau)d\tau = N^2 \tag{3.95}$$

Thus eq. (3.94) becomes:

$$x_{lf}(\tau) = \frac{M}{N} h(t-\tau) \tag{3.96}$$

The construction of $x_{lf}(\tau)$ from $h(t-\tau)$ is shown in Fig. 3.15.

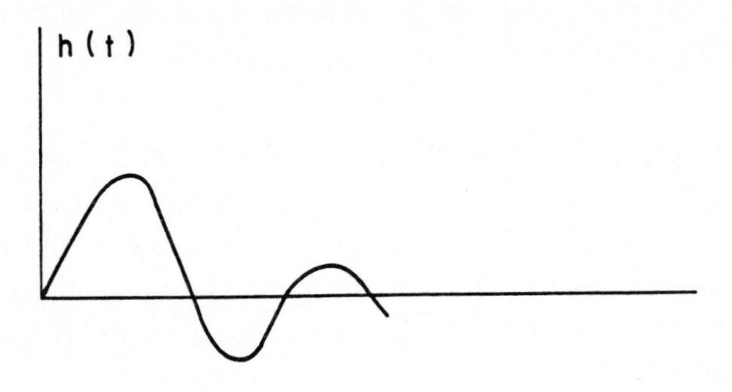

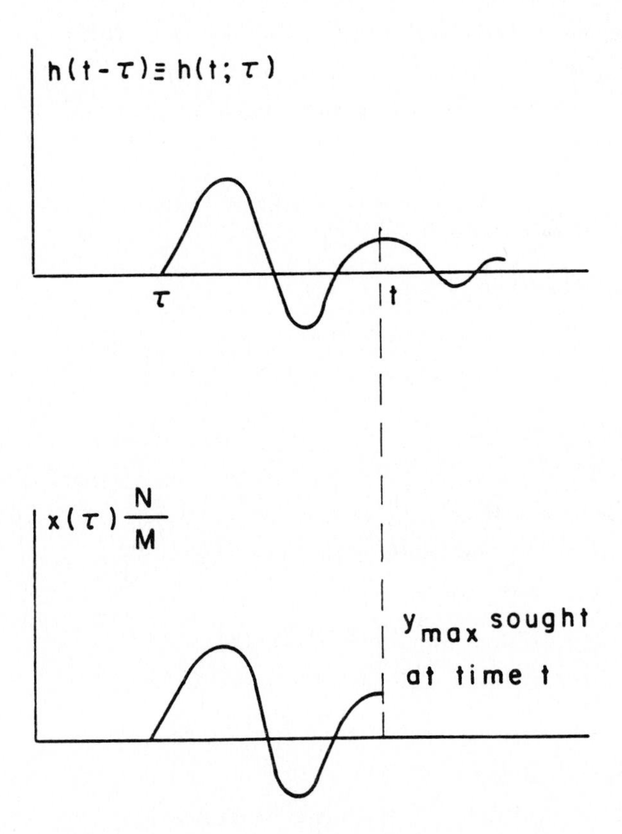

Figure 3.15: Constructing the least favorable excitation by matching the excitation to the impulse response.

The least favorable response is found by combining eqs. (3.88), (3.90) and (3.93):

$$y_{lf}(t) = MN \tag{3.97}$$

Drenick designates $x_{lf}(t)$ and $y_{lf}(t)$ as the critical excitation and critical response, respectively. The attractive feature of Drenick's approach is that it leads to results which are valid when the information on the excitation is limited to constraints of the form of eq. (3.85) and when the total energy is the only quantity whose bound is known. However, as a number of investigators note, the price paid for this paucity of information is an excessive conservatism.

On the other hand, one may be able to obtain more information than just a bound on the excitation energy. Shinozuka (1970) has suggested characterizing the excitation uncertainty by specifying an envelope of the Fourier amplitude spectrum. In order to report Shinozuka's results, we use the Fourier transform representation for the excitation as follows:

$$x(t) = \frac{1}{2\pi} \int_{-\infty}^{\infty} X(\omega) e^{i\omega t}\, d\omega \tag{3.98}$$

where $X(\omega)$ is the spectrum:

$$X(\omega) = \int_{-\infty}^{\infty} x(t) e^{-i\omega t}\, d\omega \tag{3.99}$$

According to Parseval's theorem:

$$\int_{-\infty}^{\infty} x^2(t) dt = \frac{1}{2\pi} \int_{-\infty}^{\infty} |X(\omega)|^2\, d\omega \tag{3.100}$$

Hence the energy-bound can be written as:

$$\frac{1}{2\pi} \int_{-\infty}^{\infty} |X(\omega)|^2\, d\omega \ \leq \ M^2 \tag{3.101}$$

In the frequency domain the response becomes:

$$y(t) = \frac{1}{2\pi} \int_{-\infty}^{\infty} H(\omega) X(\omega) e^{i\omega t} d\omega \tag{3.102}$$

where $H(\omega)$ is the frequency response function of the system.

Now, let us note that:

$$N^2 = \int_{-\infty}^{\infty} h^2(t)dt = \frac{1}{2\pi}\int_{-\infty}^{\infty} |H(\omega)|^2 \, d\omega \tag{3.103}$$

With the Cauchy-Schwarz inequality and relation (3.101), the left-hand side of eq. (3.102) can be expressed as:

$$|y(t)| = \frac{1}{2\pi}\int_{-\infty}^{\infty} |X(\omega)| \, |H(\omega)| \, d\omega \tag{3.104}$$

$$\leq \frac{1}{2\pi}\sqrt{\int_{-\infty}^{\infty} |X(\omega)|^2 \, d\omega}\sqrt{\int_{-\infty}^{\infty} |H(\omega)|^2 \, d\omega} \tag{3.105}$$

$$\leq MN \tag{3.106}$$

This becomes Drenick's classical result, given by eq. (3.97), when $X(\omega)$ is proportional to $H(\omega)$. Now, following Shinozuka, we assume that we have knowledge of additional constraints on the excitation spectrum. In particular, we presume knowledge of an envelope on the magnitude of the Fourier spectrum of the excitation:

$$|X(\omega)| \leq X_e(\omega) \tag{3.107}$$

Then, in view of relation (3.104), we have:

$$|y(t)| \leq \frac{1}{2\pi}\int_{-\infty}^{\infty} |H(\omega)| \, |X(\omega)| \, d\omega$$

$$\leq \frac{1}{2\pi}\int_{-\infty}^{\infty} |H(\omega)| \, X_e(\omega) \, d\omega \tag{3.108}$$

Consider the particular example of a single degree of freedom system governed by the differential equation:

$$\frac{d^2y}{dt^2} + 2\omega_0\zeta\frac{dy}{dt} + \omega_0^2 y = x(t) \tag{3.109}$$

The impulse response function $h(t)$ and the frequency response function $H(\omega)$ are given by

$$h(t) = \left[e^{-\zeta\omega_0 t}\sin\omega_d t/\omega_d\right]U(t) \tag{3.110}$$

$$H(\omega) = \frac{1}{\omega_0^2 - \omega^2 + 2i\zeta\omega_0\omega} \tag{3.111}$$

where

$$\omega_d = \omega_0\sqrt{1-\zeta^2} \tag{3.112}$$

and $U(t)$ is the Heaviside unit step function.

Shinozuka (1970) proposed three analytical forms for the envelope X_e, as follows:

$$X_{e1}(\omega) = \sigma_0\sqrt{\frac{1+4\zeta_g^2(\omega/\omega_g)^2}{[1-(\omega/\omega_g)^2]^2+4\zeta_g^2(\omega/\omega_g)^2}} \tag{3.113}$$

$$X_{e2}(\omega) = \sigma_0\frac{1}{\sqrt{[1-(\omega/\omega_g)^2]^2+4\zeta_g^2(\omega/\omega_g)^2}} \tag{3.114}$$

$$X_{e3}(\omega) = \sigma_0\left(\frac{\omega}{\omega_g}\right)^2\exp\left[-\left(\frac{\omega}{\omega_g}\right)^2\right] \tag{3.115}$$

where σ_0, ζ_g and ω_g are positive parameters.

It is instructive to compare the response-bounds of the Drenick and Shinozuka models. Drenick's model of uncertainty in the excitation function is specified by the single parameter M, which bounds the integral square value of $x(t)$. Parseval's theorem then implies relation (3.101). Drenick's model of the excitation uncertainty imposes no constraints on the form of $X(\omega)$ other than eq. (3.101). The Shinozuka model, on the other hand, restricts $X(\omega)$ to an envelope. In order to compare these models on an equal basis, let us evaluate Drenick's bound with M chosen as:

$$M_{ei}^2 = \frac{1}{2\pi}\int_{-\infty}^{\infty} X_{ei}^2(\omega)\,d\omega \quad , \quad i = 1,2,3 \tag{3.116}$$

The value of N is obtained from eqs. (3.95) and (3.110). Then NM_{ei}, $i = 1,2,3$, are Drenick's values of the response bounds corresponding to

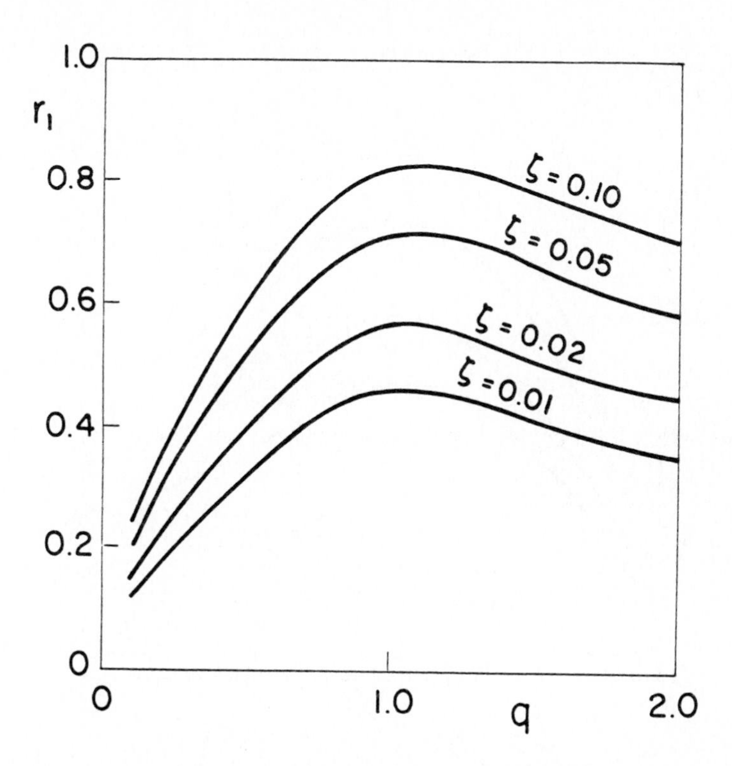

Figure 3.16: Response-bound ratio r_1 based on excitation envelope $X_{e1}(\omega)$. (After Shinozuka, 1970. Reproduced by permission of the American Society of Civil Engineers.)

each of Shinozuka's envelopes. The corresponding response bounds in Shinozuka's model are:

$$I_{ei} = \frac{1}{2\pi} \int_{-\infty}^{\infty} |H(\omega)| \, X_{ei}(\omega) \, d\omega \quad , \quad i = 1, 2, 3 \tag{3.117}$$

where $H(\omega)$ is given by eq. (3.111).

Following Shinozuka, we consider the ratio of the Shinozuka response-bounds to the Drenick response-bounds:

$$r_i = \frac{I_{ei}}{N M_{ei}} \quad , \quad i = 1, 2, 3 \tag{3.118}$$

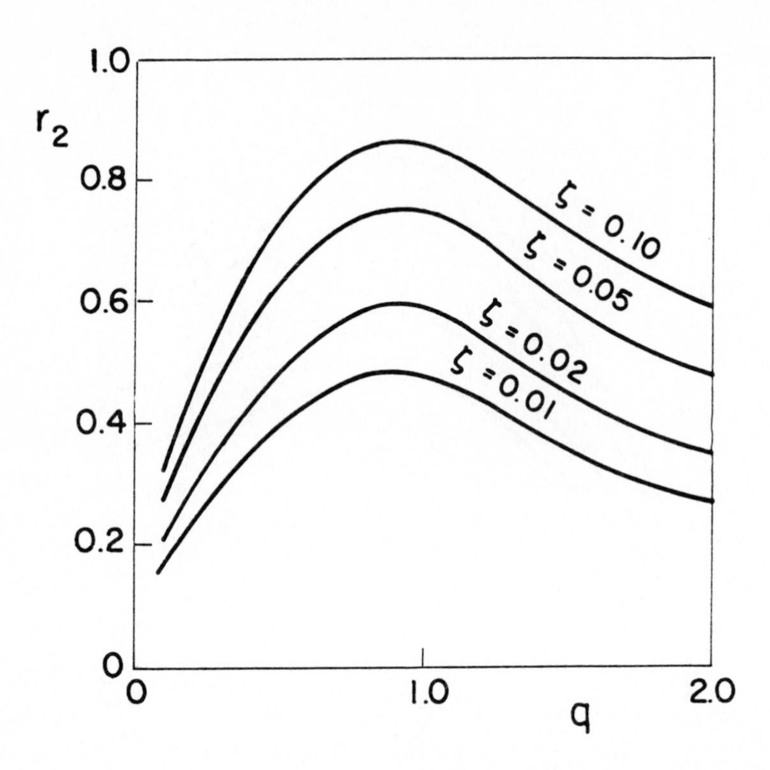

Figure 3.17: Response-bound ratio r_2 based on excitation envelope $X_{e2}(\omega)$. (After Shinozuka, 1970. Reproduced by permission of the American Society of Civil Engineers.)

Figs. 3.16 – 3.18 depict r_1, r_2 and r_3, respectively, as functions of the non-dimensional frequency:

$$q = \frac{\omega_0}{\omega_g} \tag{3.119}$$

Various values of the damping parameter, ζ, are shown. Here ω_g was fixed at 5 cps and ζ_g at 0.5. It should be noted that the values r_1 and r_2 are much larger than those of r_3 when $q \ll 1$ or $\omega_0 \ll \omega_g$. This phenomenon is explained by the fact that $X_{e1}(\omega)$ and $X_{e2}(\omega)$ approach unity whereas $X_{e3}(\omega)$ tends to zero as the frequency ω increases.

The most important feature illustrated by Figs. 3.16 – 3.18 is that r_i is invariably less than unity. This means that Shinozuka's model

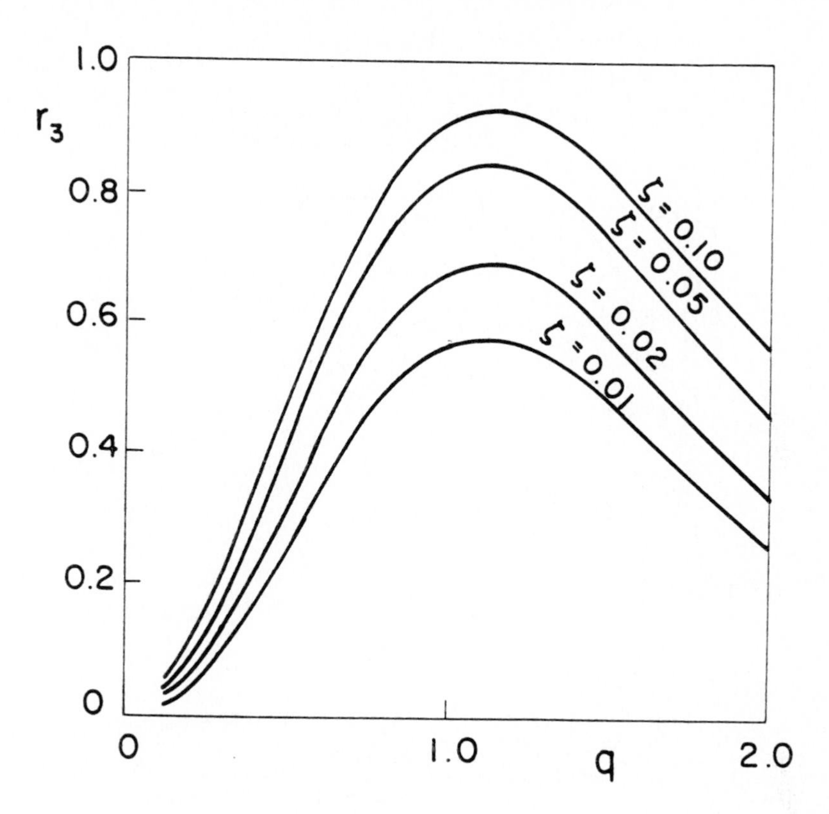

Figure 3.18: Response-bound ratio r_3 based on excitation envelope $X_{e3}(\omega)$. (After Shinozuka, 1970. Reproduced by permission of the American Society of Civil Engineers.)

provides a lower estimate of the maximum response than provided by Drenick's model. That is, the information embodied in Shinozuka's envelopes on $X(\omega)$ constrains the set of possible excitation functions and correspondingly reduces the magnitude of the response as well. In practical terms, knowledge of the excitation envelope produces less conservative estimates of the structural response. This relaxation of Drenick's more conservative estimate is acceptable if the excitation envelopes are reliable.

Inspection of Figs. 3.16 – 3.18 reveals that the most substantial reduction in the maximum response is achieved for those structures with values of q different from unity. This is explained by the fact that

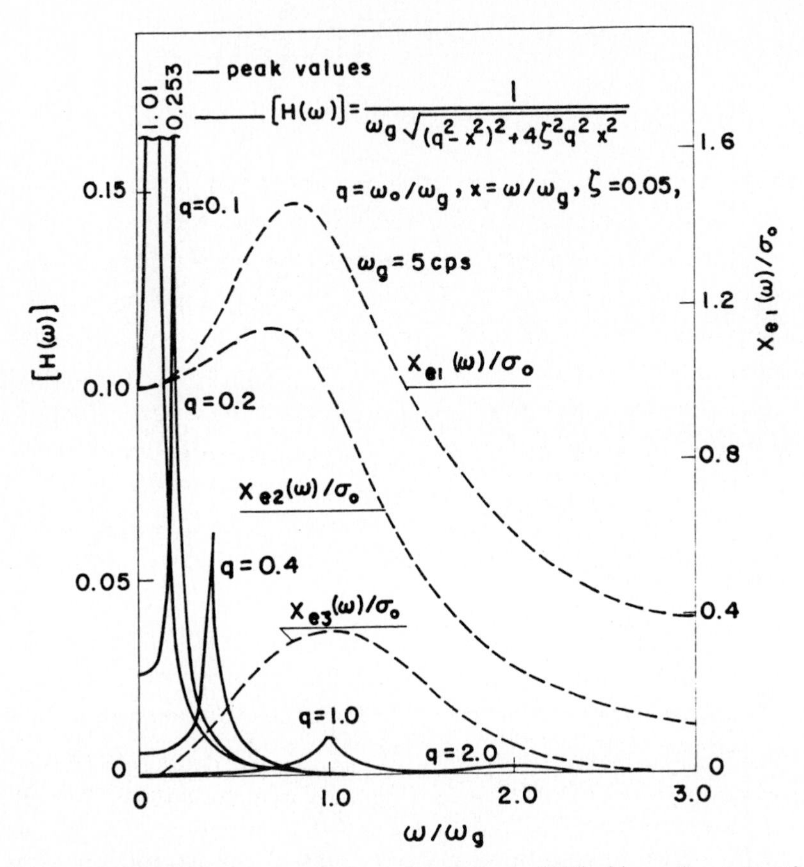

Figure 3.19: Frequency response functions $|H(\omega)|$ and excitation envelopes $X_{ei}(\omega)$. (After Shinozuka, 1970. Reproduced by permission of the American Society of Civil Engineers.)

$X_{ei}(\omega)$ and $|H(\omega)|$ are roughly proportional when $q \approx 1$ (see Fig. 3.19).

These results have been extended in several directions. Iyengar (1970) employed nonstationary excitation envelopes in the same approach. Wang *et al* (1978) introduced the concept of subcritical excitation, which enhances the practical applicability of Drenick's least favorable excitation concept. This excitation was so chosen that it differs from a set of recorded accelerograms in a minimum square sense. The method has been extended for the case of multiple constraints, with additional constraints on velocity and acceleration, by Drenick and his associates (Drenick *et al* 1979, 1984; Bedrosian *et al* 1980).

3.4 Vibration Measurements

3.4.1 Introduction

The measurement of vibrations in mechanical systems has been used extensively in many areas of industry to monitor structural or operational integrity. Rotating machinery generates particularly simple vibrational power spectra and has been the subject of extensive performance monitoring by vibrational analysis (Ricca and Bradshaw, 1984; Laws and Muszynska, 1987; Hessel *et al*, 1988). The vibrational power spectra of reciprocating machines are more complicated than for rotating machinery, but high-speed digital data acquisition systems enable vibrational monitoring of piston-cylinder performance (Goldman, 1987). Elastic vibrations are monitored in many civil structures and other mechanical systems, including oil rigs (Basseville, *et al*, 1986; Shahrivar and Bouwkamp, 1986), heat-exchanger tubing (Jendrzejczyk, 1986) and condensate piping (Ozol, 1987) and radar towers (Hoff and Natke, 1989).

An extensive array of instrumentation is available for performing vibration measurements. Transducers for recording displacement, velocity or acceleration are available commercially, covering a wide range of frequencies and amplitudes (White, 1985; Stumpf, 1987; Laws and Muszynska, 1987). In some circumstances the mechanical vibration generates temporal variation in other properties of the system. An outstanding example is the monitoring of the mechanical integrity of nuclear-reactor core components by measuring vibration-induced neutron noise (Williams, 1974; Thie, 1981; Akerhielm, Espefalt and Lorenzen, 1982).

A challenge common to all vibrational monitoring applications is the proper positioning of sensors. Monitoring of machine or structural integrity is usually a task in remote sensing, requiring signal analysis to discern changes in components which are not accessible for direct inspection. Several sources of uncertainty must be overcome in this analysis. Foremost is the wide range of pathological conditions which must be diagnosed. The uncertainty in the precise form of the malfunction is both temporal and spatial, since mechanical failures can develop in time (for which early detection is often crucial) and can occur in a range of geometrical configurations in the system being monitored. For example, flow blockage in heat-exchanger tubing can evolve gradually in time and can develop in an endless gradation of spatial occurrences. An additional source of uncertainty arises from auxiliary random phenomena which interfer with the detection of system changes. An example is provided by marine growth on submarine structural elements of off-shore oil rigs, which interferes with the vibrational monitoring of the structural integrity of the rig. Finally, vibrational monitoring, when used for system identification or for planning preventive maintenance, employs models of dynamical and physical properties of the system. Uncertainty in these models reduces the diagnostic power of the vibration measurements. In short, three major classes of uncertainty complicate the design of a vibration measurement: uncertainty in the abnormal conditions to be diagnosed, auxiliary random phenomena and model uncertainties. The selection and deployment of the vibration sensors must provide robustness of the diagnosis with respect to all these sources of uncertainty.

The task of optimally locating sensors in distributed systems has been the subject of extensive research based on probabilistic techniques (Kubrusly and Malebranche, 1985). However, the spatial uncertainty which is characteristic of failures and of auxiliary random phenomena in mechanical systems presents special difficulties. The tremendous complexity of these uncertain phenomena make probabilistic models difficult to implement. In this section it will be seen that convex models of uncertainty can be employed to provide meaningful design information, while requiring only modest amounts of data on the sources of uncertainty.

3.4.2 Damped Vibrations: Full Measurement

Consider a multi-dimensional elastic structure subject to small vibrations. Each vibrational mode of the structure is described by a linear differential equation. If the order of any of these differential equations is greater than one, then that equation can be replaced by a set of linear differential equations of first order by introducing additional state variables which are defined as the derivatives of existing state variables. In this way a set of first-order differential equations is obtained which describes the dynamic behaviour of the structure.

In normal operation the system is excited by external forces which induce vibrations, and the state variables are measured at a certain instant after initiation of the excitation. These normal excitations may be accompanied by uncertainty, so we will represent the normal excitations with a set of normal excitation functions. Furthermore, many abnormalities of the system (as opposed to nominal uncertainty) are manifested in alteration of the set of excitation functions. Each type of abnormality corresponds to a distinct set of excitation functions. Diagnosis of system malfunctions requires differentiating between sets of excitations on the basis of measurments of the system response. The aim of the analysis in this section is to establish what sets of excitations can be distinguished on the basis of these measurements. In particular, it is desired to distinguish between normal and abnormal excitation sets, and between various sets of abnormal excitations.

Let $x^T(t) = (x_1(t), \ldots, x_N(t))$ represent the N vibrational state variables at time t. Similarly, let $f^T(t) = (f_1(t), \ldots, f_N(t))$ represent the vector excitation function. $f(t)$ belongs to a set of excitation functions: either the nominal range of excitations or a set of abnormal disturbances. The dynamics of the system are described by:

$$\frac{dx}{dt} = Kx(t) + f(t) \tag{3.120}$$

where K is a constant $N \times N$matrix. Assuming that the initial values of the state variables are zero, the response of the system to vector excitation $f(t)$ is:

$$x_f(t) = \int_0^t e^{K(t-\tau)} f(\tau)\, d\tau \tag{3.121}$$

We shall assume that the excitation function of each mode is uniformly bounded in time. Thus the sets of vector excitation functions are of the following form:

$$F_{a,b} = \left\{ f^T = (f_1, \ldots, f_N) : \; a_n \le f_n(t) \le b_n \quad , \quad n = 1, \ldots, N \right\} \tag{3.122}$$

The components of the nominal excitation functions may vary around zero. For instance, the bounds of variation of the nominal excitation may satisfy:

$$a = -b \tag{3.123}$$

Abnormal conditions are characterized by extension of the range of variation of the excitation and/or shift of the center of the range. Abnormalities may develop in some or in all of the components of the excitation function.

Each element of the excitation set $F_{a,b}$ produces a response x_f according to eq. (3.121). The set of all responses to excitations in the set $F_{a,b}$ is the complete response set $C_{a,b}$, defined as:

$$C_{a,b} = \{x : x = x_f(t) \quad \text{for all} \quad f \in F_{a,b}\} \tag{3.124}$$

The goal of the analysis is to determine which excitation sets can be distinguished from one another on the basis of measurement of the response vector at a given instant after initiation of the excitation. Two excitation sets $F_{a,b}$ and $F_{\alpha,\beta}$ are always distinguishable if and only if the corresponding response sets $C_{a,b}$ and $C_{\alpha,\beta}$ are disjoint. That is, a single measurement of the response vector can differentiate between every excitation vector bounded by the parameters $(a_1, b_1, \ldots, a_N, b_N)$ and every excitation bounded by the parameters $(\alpha_1, \beta_1, \ldots, \alpha_N, \beta_N)$ if and only if:

$$C_{a,b} \cap C_{\alpha,\beta} = \emptyset \tag{3.125}$$

The excitation sets defined by eq. (3.122) are convex and the system response is a linear function of the excitation as seen in eq. (3.121). A linear transformation of a convex set produces a convex set, so the response sets are convex. Consequently, the concept of hyperplane separation, described in section 2.3.3, can be used to establish conditions for the disjointness of the response sets. Specifically, eq. (3.125) holds if and only if there exists a vector ω such that:

$$\max_{c \in C_{a,b}} \omega^T c < \min_{d \in C_{\alpha,\beta}} \omega^T d \tag{3.126}$$

This relation states that the convex response sets $C_{a,b}$ and $C_{\alpha,\beta}$ are disjoint if and only if there exists a direction ω in measurement space such that the "highest" plane perpendicular to ω and intersecting $C_{a,b}$ is "below" the "lowest" plane perpendicular to ω and intersecting $C_{\alpha,\beta}$.

Each element of the response set $C_{a,b}$ or $C_{\alpha,\beta}$ is the transformation of an element of the corresponding excitation set, so relation (3.126) is equivalent to:

$$\max_{f\in F_{a,b}} \omega^T x_f \; < \; \min_{g\in F_{\alpha,\beta}} \omega^T x_g \tag{3.127}$$

We now wish to develop expressions for these extrema. Let $\gamma^n(t)$ represent the nth column of the $N\times N$ matrix e^{Kt}. Employing eq. (3.121) we find that $\omega^T x_f$ can be written as:

$$\omega^T x_f = \sum_{n=1}^{N} \int_0^t \omega^T \gamma^n(t-\tau) f_n(\tau)\, d\tau \tag{3.128}$$

Examination of this equation reveals that $\omega^T x_f$ is maximized on $F_{a,b}$ by choosing $f_n(\tau) = b_n$ when $\omega^T \gamma^n(t-\tau)$ is non-negative and by choosing $f_n(\tau) = a_n$ when $\omega^T\gamma^n(t-\tau)$ is negative, for each $n = 1,\ldots,N$. Let D_{n+} denote the set of values of τ in $[0,t]$ for which $\omega^T\gamma^n(t-\tau)$ is non-negative, while D_{n-} is the complement of D_{n+} in $[0,t]$. That is,

$$D_{n+} = \left\{\tau:\; 0 \le \tau \le t \;\;,\;\; \omega^T\gamma^n(t-\tau) \ge 0\right\} \tag{3.129}$$

$$D_{n-} = \left\{\tau:\; 0 \le \tau \le t \;\;,\;\; \omega^T\gamma^n(t-\tau) < 0\right\} \tag{3.130}$$

Thus the maximum of $\omega^T x_f$ becomes

$$\max_{f\in F_{a,b}} \omega^T x_f = \sum_{n=1}^{N} \left(b_n \int_{D_{n+}} \omega^T\gamma^n(t-\tau)\, d\tau + a_n \int_{D_{n-}} \omega^T\gamma^n(t-\tau)\, d\tau \right) \tag{3.131}$$

Again referring to eq. (3.128) one finds that $\omega^T x_g$ is minimized on $F_{\alpha,\beta}$ by choosing $g_n(\tau) = \alpha_n$ when $\omega^T\gamma^n(t-\tau)$ is non-negative and by choosing $g_n(\tau) = \beta_n$ when $\omega^T\gamma^n(t-\tau)$ is negative. Thus

$$\min_{g\in F_{\alpha,\beta}} \omega^T x_g = \sum_{n=1}^{N} \left(\alpha_n \int_{D_{n+}} \omega^T\gamma^n(t-\tau)\, d\tau + \beta_n \int_{D_{n-}} \omega^T\gamma^n(t-\tau)\, d\tau \right) \tag{3.132}$$

3.4.3 Example: 2-Dimensional Measurement

Let us consider a numerical example in order to clarify the application of these equations to the determination of the distinguishability of excitation sets. Consider a 2-dimensional system for which the matrix K is

$$K = \begin{pmatrix} -1 & -1 \\ -1 & -1 \end{pmatrix} \tag{3.133}$$

Each vibrational mode has a self-restoring force proportional to its own displacement and a restoring force proportional to the displacement of the other mode.

Elementary matrix calculations lead to the representation of e^{Kt} as

$$e^{Kt} = \frac{1}{2} \begin{pmatrix} e^{-2t} + 1 & e^{-2t} - 1 \\ \\ e^{-2t} - 1 & e^{-2t} + 1 \end{pmatrix} \tag{3.134}$$

The columns of this matrix are the vectors $\gamma^1(t)$ and $\gamma^2(t)$.

Let $F_{a,b}$ be the set of excitation functions whose elements are bounded by ± 1. Thus $(a_i, b_i) = (-1, 1)$ for $i = 1, 2$. This set can be represented by a 2×2 square centered at the origin in the f_2–versus–f_1 plane, as shown in fig. 3.20: The axes indicate the range of variation of the components of the excitation function $f(t)$. Let $F_{\alpha,\beta}$ be a different excitation set, where $\beta_1 = \alpha_1 + 1.5$ and $\beta_2 = \alpha_2 + 1.0$. Thus $F_{\alpha,\beta}$ can be represented by a rectangle, as in fig. 3.20 for $\alpha_1 = -0.5$ and $\alpha_2 = 1.5$.

Keeping the set $F_{a,b}$ fixed, we will enquire what sets $F_{\alpha,\beta}$ can be distinguished from the set $F_{a,b}$ on the basis of a single measurement of the vector response of the system at time $t = 1$. Specifically, for a given choice of α_1 and α_2, we will determine whether the resulting response set $C_{\alpha,\beta}$ is disjoint from the response set $C_{a,b}$.

The response to an excitation function is a 2-component vector, (c_1, c_2). Figures 3.21 to 3.23 show the response sets $C_{a,b}$ and $C_{\alpha,\beta}$ for three different choices of the α_1 and α_2. Figure 3.21 shows response sets which are disjoint. This means that every realization of the corresponding excitation functions are distinguishable on the basis of a single measurement of the system. $C_{\alpha,\beta}$ approaches $C_{a,b}$ as the values

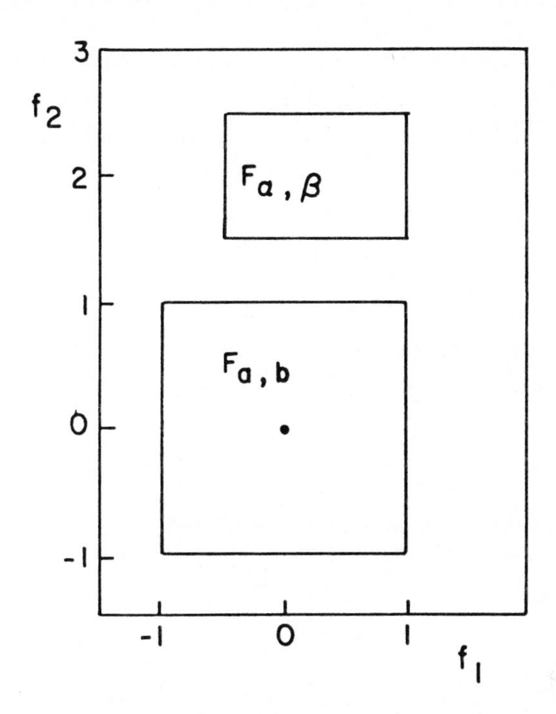

Figure 3.20: Graphical representation of uniform bounds on the excitation sets $F_{a,b}$ and $F_{\alpha,\beta}$.

of (α_2, β_2) are reduced, as seen by comparing figs. 3.21 and 3.22. Further reduction of (α_2, β_2), as in fig. 3.23, causes the response sets to overlap, which means that not all excitation functions in $F_{a,b}$ are distinguishable from all the excitation functions in $F_{\alpha,\beta}$, despite the fact that these excitation sets are themselves disjoint.

The systematic determination of the values of (α_1, α_2) for which the excitation sets $F_{a,b}$ and $F_{\alpha,\beta}$ are always distinguishable depends on the numerical determination of the disjointness of the corresponding response sets. $F_{a,b}$ and $F_{\alpha,\beta}$ are always distinguishable if and only if there exists a vector ω which satisfies relation (3.127). This is implemented numerically for a specific choice of (α_1, α_2) by a 2–loop iteration. In the outer loop the vector ω is varied, while in the inner loop the extrema in eqs. (3.131) and (3.132) are employed to determine if (3.127) is satisfied

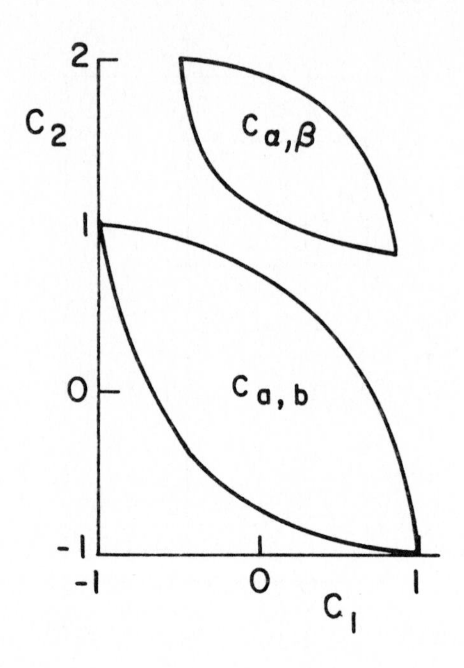

Figure 3.21: Response sets $C_{a,b}$ and $C_{\alpha,\beta}$. $(\alpha_1, \beta_1) = (0.5, 2.0)$; $(\alpha_2, \beta_2) = (2.0, 3.0)$.

by the current value of ω.

Typical results of this analysis are shown in fig. 3.24. Each point on this α_2–versus–α_1 plane represents a choice of an excitation set $F_{\alpha,\beta}$ for which $\beta_1 = \alpha_1 + 1.5$ and $\beta_2 = \alpha_2 + 1.0$. Thus for example the point $(\alpha_1, \alpha_2) = (-0.5, 1.2)$ represents the set $F_{\alpha,\beta}$ for which $(\alpha_1, \beta_1) = (-0.5, 1.0)$ and $(\alpha_2, \beta_2) = (1.2, 2.2)$. All the excitation sets shown in the figure are in fact disjoint from $F_{a,b}$ (represented by the 2×2 square in fig. 3.20.) However, only those sets represented by points above the curve generate response sets $C_{\alpha,\beta}$ which are disjoint from the response set $C_{a,b}$. Each point below the curve in fig. 3.24 corresponds to an excitation set which contains elements that are indistinguishable (on the basis of a single vector measurement of the system at time $t = 1$) from some excitation in $F_{a,b}$.

We have now achieved our aim of establishing which excitation sets $F_{\alpha,\beta}$ are distinguishable from $F_{a,b}$ and which are not, on the basis of a

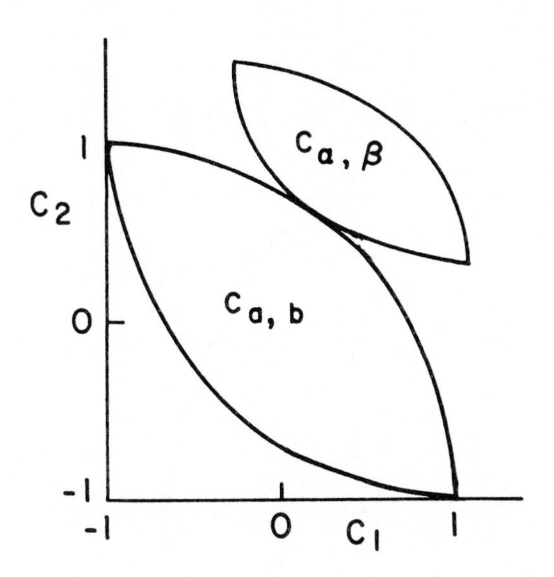

Figure 3.22: Response sets $C_{a,b}$ and $C_{\alpha,\beta}$. $(\alpha_1, \beta_1) = (0.5, 2.0)$; $(\alpha_2, \beta_2) = (1.225, 2.225)$.

single vector measurement at time $t = 1$.

3.4.4 Damped Vibrations: Partial Measurement

In the previous two subsections we assumed that each vibrational mode of the system is directly measured. This is not always realistic, especially for systems of high dimensionality. In this subsection we consider the selection of M measurements from among the N vibrational modes, where $M \leq N$. Eq. (3.120) describes the dynamics of the system. However, the measurement vector y_f in response to excitation $f(t)$ is now a linear combination of the state variables:

$$y_f = Gx_f \tag{3.135}$$

In general G is an arbitrary $M \times N$ matrix. However, some special cases are of particular interest. If each mode is measured directly then G is

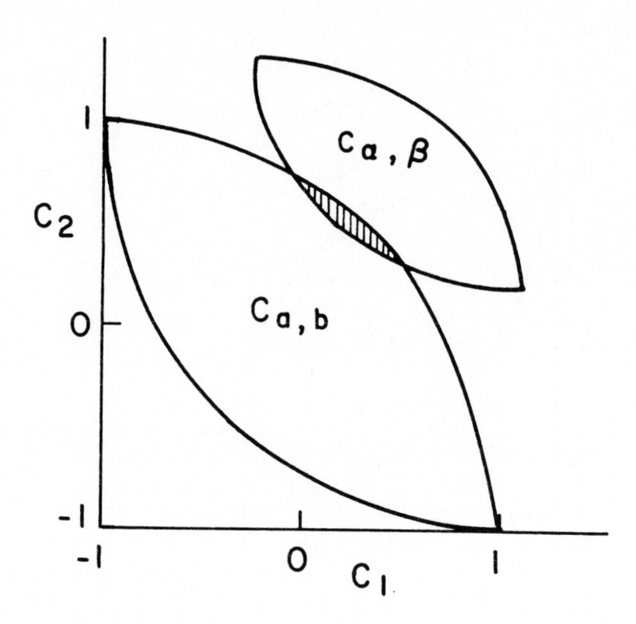

Figure 3.23: Response sets $C_{a,b}$ and $C_{\alpha,\beta}$. $(\alpha_1, \beta_1) = (0.5, 2.0)$; $(\alpha_2, \beta_2) = (1.05, 2.05)$.

the $N \times N$ identity matrix. If selected modes are measured directly and other modes not measured at all, then G is obtained from the $N \times N$ identity matrix by removing the rows corresponding to unmeasured vibrational modes.

The system is excited by uniformly bounded excitations, so the excitation sets are specified as in eq. (3.122). The set of all M-vector measurements realizable in response to excitation vectors in $F_{a,b}$ is now:

$$C_{a,b} = \{y : \ y = Gx_f \quad \text{for all} \quad f \in F_{a,b}\} \tag{3.136}$$

A selection of measurements specified by the measurement matrix G distinguishes between excitation sets $F_{a,b}$ and $F_{\alpha,\beta}$ if and only if the corresponding response sets are disjoint. Response sets $C_{a,b}$ and $C_{\alpha,\beta}$ are disjoint if and only if there exists an M-vector μ such that:

$$\max_{c \in C_{a,b}} \mu^T c < \min_{d \in C_{\alpha,\beta}} \mu^T d \tag{3.137}$$

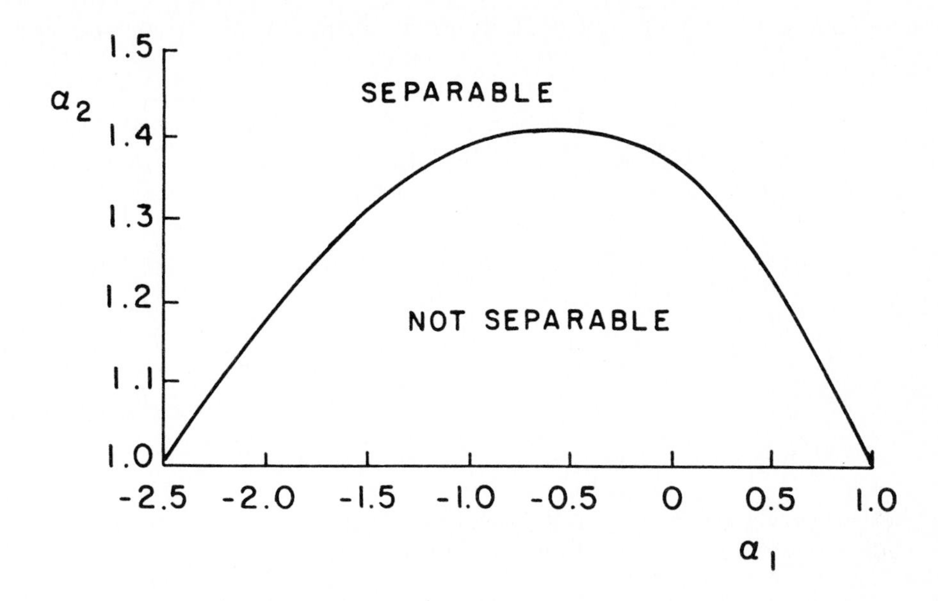

Figure 3.24: Regions of separable and not separable response sets.

This relation is equivalent to:

$$\max_{f \in F_{a,b}} \mu^T G x_f < \min_{g \in F_{\alpha,\beta}} \mu^T G x_g \tag{3.138}$$

Let us compare this with relation (3.127). If there is an M-vector μ satisfying (3.138), then $\omega = G^T \mu$ is an N-vector satisfying (3.127). Thus if M measurements are sufficient to differentiate between the excitation sets $F_{a,b}$ and $F_{\alpha,\beta}$ then $N(\geq M)$ measurements are also sufficient. On the other hand, if no N-vector ω exists which satisfies (3.127), then it is clear that there is no M-vector μ satisfying (3.138), for if such a μ existed, then $\omega = G^T \mu$ would satisfy (3.127). In other words, if complete measurement of all N modes is insufficient to distinguish between excitation sets $F_{a,b}$ and $F_{\alpha,\beta}$, then partial measurement is surely insufficient. However, an M-vector satisfying (3.138) does not necessarily exist even if an N-vector exists which satisfies (3.127). This means that a subset of M measurements may not suffice to distinguish $F_{a,b}$ from $F_{\alpha,\beta}$, even though all N measurements are sufficient. The aim of the present analysis is to evaluate the degradation of the resolution between excitation sets as the selection of measured variables is restricted.

As before, let $\gamma^n(t)$ represent the nth column of e^{Kt}. Then $\mu^T G x_f$ can be written as

$$\mu^T G x_f = \sum_{n=1}^{N} \int_0^t \mu^T G \gamma^n(t-\tau) f_n(\tau)\, d\tau \tag{3.139}$$

Now, in analogy to eqs. (3.129) and (3.130), let us define the following sets:

$$D_{n+} = \left\{ \tau : \ 0 \le \tau \le t \ , \ \mu^T G \gamma^n(t-\tau) \ge 0 \right\} \tag{3.140}$$

$$D_{n-} = \left\{ \tau : \ 0 \le \tau \le t \ , \ \mu^T G \gamma^n(t-\tau) < 0 \right\} \tag{3.141}$$

Thus the extrema in eq. (3.138) become:

$$\max_{f \in F_{a,b}} \mu^T G x_f = \sum_{n=1}^{N} \left(b_n \int_{D_{n+}} \mu^T G \gamma^n(t-\tau) d\tau + a_n \int_{D_{n-}} \mu^T G \gamma^n(t-\tau) d\tau \right) \tag{3.142}$$

$$\min_{g \in F_{\alpha,\beta}} \mu^T G x_g = \sum_{n=1}^{N} \left(\alpha_n \int_{D_{n+}} \mu^T G \gamma^n(t-\tau) d\tau + \beta_n \int_{D_{n-}} \mu^T G \gamma^n(t-\tau) d\tau \right) \tag{3.143}$$

The excitation sets $F_{a,b}$ and $F_{\alpha,\beta}$ are always distinguishable with the M measurements determined by the measurement matrix G if and only if there exists an M-vector μ such that relation (3.138) holds. Eqs. (3.142) and (3.143) provide explicit expressions for the extrema involved.

3.4.5 Transient Vibrational Acceleration

Sections 3.4.2 to 3.4.4 are devoted to the optimization of vibrational measurements of uniformly bounded excitations. Such excitation functions represent continual or recurrent bounded disturbances to the system. Unbounded but transient disturbances are not conveniently represented by the uniform-bound convex model. The diagnosis of transient disturbances is important for early detection of failure and for detection of infrequent abnormal conditions.

A useful representation of transient disturbances is the energy-bound convex model. Consider the following set of vector-valued functions:

$$F(\bar{f},E)=\left\{f(t):\int_0^{T_f}\left(f(t)-\bar{f}(t)\right)^T\left(f(t)-\bar{f}(t)\right)\,dt\le E\right\} \tag{3.144}$$

where $\bar{f}(t)$ is a specified vector function and E is a positive number. $F(\bar{f},E)$ contains all the functions whose integral squared deviation from $\bar{f}(t)$ does not exceed E. We will say that the "energy" of deviation from $\bar{f}$ of elements of $F(\bar{f},E)$ does not exceed E. An element of $F(\bar{f},E)$ may deviate from $\bar{f}$ by a large amount, but only for a limited duration. In other words, elements of $F(\bar{f},E)$ represent transient deviations from $\bar{f}$ of bounded energy and unbounded magnitude.

$F(\bar{f},E)$ is a convex set, and its extreme-point functions are the functions whose integral energy of deviation precisely equals E. Consequently, $F(\bar{f},E)$ is the convex hull of the set:

$$H(\bar{f},E)=\left\{f:\int_0^{T_f}\left(f(t)-\bar{f}(t)\right)^T\left(f(t)-\bar{f}(t)\right)\,dt=E\right\} \tag{3.145}$$

Consider the dynamic system described by eq. (3.120), where K is a constant $N\times N$ matrix. The response to excitation functions $f(t)$ is $x_f(t)$, defined in eq. (3.121). The measurement vector in response to excitation function $f(t)$ is $y_f(t)$, defined in eq. (3.135). y_f is an M-vector, where $M\le N$, and G is a constant $M\times N$ matrix. The set of all measurement vectors at time T_f for excitations in $F(\bar{f},E)$ is the response set, $C(\bar{f},E)$:

$$C(\bar{f},E)=\left\{y:\ y=y_f(T_f),\quad \text{for all}\quad f\in F(\bar{f},E)\right\} \tag{3.146}$$

Let $F(\bar{f},E_1)$ and $F(\bar{g},E_2)$ be two sets of energy-bounded excitations. For example, $F(\bar{f},E_1)$ may represent the normal behaviour of the system in which $\bar{f}$ is identically zero and E_1 is small, corresponding to small perturbations near zero. $F(\bar{g},E_2)$ may represent an abnormal situation in which $\bar{g}$ differs from zero and E_2 may be large. Alternatively, these excitation sets may both represent distinct abnormal situations. In any case, the aim of our analysis is to determine if the measurements, defined by the measurement matrix G, distinguish between distinct excitation sets.

All excitations in $F(\bar{f}, E_1)$ are distinguishable from all excitations in $F(\bar{g}, E_2)$ if and only if the corresponding response sets are disjoint:

$$C(\bar{f}, E_1) \cap C(\bar{g}, E_2) = \emptyset \tag{3.147}$$

Because these sets are convex, this disjointness condition holds if and only if there is a hyperplane which separates these response sets. In other words, $C(\bar{f}, E_1)$ and $C(\bar{g}, E_2)$ are disjoint if and only if there exists an M-vector μ such that:

$$\max_{f \in F(\bar{f}, E_1)} \mu^T y_f < \min_{g \in F(\bar{g}, E_2)} \mu^T y_g \tag{3.148}$$

The extremal values in this relation occur on the sets of extreme-point functions, $H(\bar{f}, E_1)$ and $H(\bar{g}, E_2)$ respectively, because $\mu^T y_f$ is a linear function of f.

It is an elementary matter to obtain the extrema of $\mu^T y_f$ by employing the Cauchy-Schwarz inequalities. Let a and b be vectors. The Cauchy inequality states that:

$$\left(a^T b\right)^2 \le \left(a^T a\right)\left(b^T b\right) \tag{3.149}$$

Equality occurs in (3.149) if a is proportional to b. Now let $u(t)$ and $v(t)$ be scalar functions. The Schwarz inequality, which is a generalization of Cauchy's inequality, asserts:

$$\left(\int u(t)v(t)\,dt\right)^2 \le \int u^2(t)\,dt \int v^2(t)\,dt \tag{3.150}$$

Again, equality results from proportionality of $u(t)$ and $v(t)$.

Employing these relations we develop expressions for the extrema in (3.148). Combining eqs. (3.121) and (3.135) we find that:

$$\mu^T\left(y_f(T) - y_{\bar{f}}(T)\right) = \mu^T G \int_0^{T_f} e^{K(T-t)}(f(t) - \bar{f}(t))\,dt \tag{3.151}$$

For convenience of notation let us define the vector function:

$$\eta^T(t) = \mu^T G e^{K(T_f - t)} \tag{3.152}$$

The Cauchy inequality implies that, at each instant t, the integrand of eq. (3.151) satisfies:

$$\eta^T(t)\left(f(t) - \bar{f}(t)\right) \leq \sqrt{\eta^T(t)\eta(t)}\sqrt{\left(f(t) - \bar{f}(t)\right)^T\left(f(t) - \bar{f}(t)\right)} \tag{3.153}$$

Thus the lefthand side of eq. (3.151) is bounded above by:

$$\mu^T(y_f(T_f) - y_{\bar{f}}(T_f)) \leq \int_0^{T_f} \sqrt{\eta^T(t)\eta(t)} \times$$

$$\sqrt{\left(f(t) - \bar{f}(t)\right)^T\left(f(t) - \bar{f}(t)\right)}dt \tag{3.154}$$

The Schwarz inequality can be applied to the righthand side of (3.154) to yield:

$$\mu^T(y_f(T_f) - y_{\bar{f}}(T_f)) \leq \sqrt{\int_0^{T_f} \eta^T(t)\eta(t)\,dt} \times$$

$$\sqrt{\int_0^{T_f} \left(f(t) - \bar{f}(t)\right)^T\left(f(t) - \bar{f}(t)\right)dt} \tag{3.155}$$

Equality is obtained in (3.153) if the vector $f(t) - \bar{f}(t)$ is proportional to the vector $\eta(t)$. Similarly, equality occurs in (3.155) if the scalar function $\left(f(t) - \bar{f}(t)\right)^T\left(f(t) - \bar{f}(t)\right)$ is proportional to the scalar function $\eta^T(t)\eta(t)$. It is evident that f can be chosen from $H(\bar{f}, E_1)$ to satisfy these conditions by choosing:

$$f(t) - \bar{f}(t) = \gamma\eta(t) \tag{3.156}$$

where γ is:

$$\gamma^2 = \frac{E_1}{\int_0^{T_f} \eta^T(t)\eta(t)\,dt} \tag{3.157}$$

Thus, f can be chosen from $H(\bar{f}, E_1)$ so that $\mu^T(y_f(T) - y_{\bar{f}})$ is in fact as large as the righthand side of (3.155). However, the second integral on the righthand side of eq. (3.155) is precisely equal to E_1. Consequently, we find that the greatest possible value of $\mu^T(y_f - y_{\bar{f}})$ is:

$$\max_{f \in H(\bar{f}, E_1)} \mu^T(y_f(T_f) - y_{\bar{f}}(T_f)) = \sqrt{E_1 \int_0^{T_f} \eta^T(t)\eta(t)\, dt} \tag{3.158}$$

A similar argument can be presented to find the minimum of $\mu^T(y_g - y_{\bar{g}})$:

$$\min_{f \in H(\bar{g}, E_2)} \mu^T(y_g(T_f) - y_{\bar{g}}(T_f)) = -\sqrt{E_2 \int_0^{T_f} \eta^T(t)\eta(t)\, dt} \tag{3.159}$$

Returning to (3.148), we find that the excitation sets $F(\bar{f}, E_1)$ and $F(\bar{g}, E_2)$ are always distinguishable if and only if there exists a vector μ such that:

$$\mu^T y_{\bar{f}}(T) + \sqrt{E_1 \int_0^{T_f} \eta^T(t)\eta(t)\, dt} < \mu^T y_{\bar{g}}(T) - \sqrt{E_2 \int_0^{T_f} \eta^T(t)\eta(t)\, dt} \tag{3.160}$$

A simple example will illuminate the meaning and utility of this result. Consider a 1-dimensional system, so K is a scalar and $N = M = 1$. μ is now a scalar and can be divided out of (3.160). Let $F(\bar{f}, E_1)$ represent the normal behaviour of the system, so:

$$\bar{f}(t) = 0 \tag{3.161}$$

Let $F(\bar{g}, E_2)$ represent abnormal disturbances around a constant excitation:

$$\bar{g}(t) = \text{constant} \tag{3.162}$$

Relation (3.160) now becomes:

$$\sqrt{E_1} + \sqrt{E_2} < \bar{g}\sqrt{\frac{2}{K}\frac{e^{KT_f} - 1}{e^{KT_f} + 1}} \tag{3.163}$$

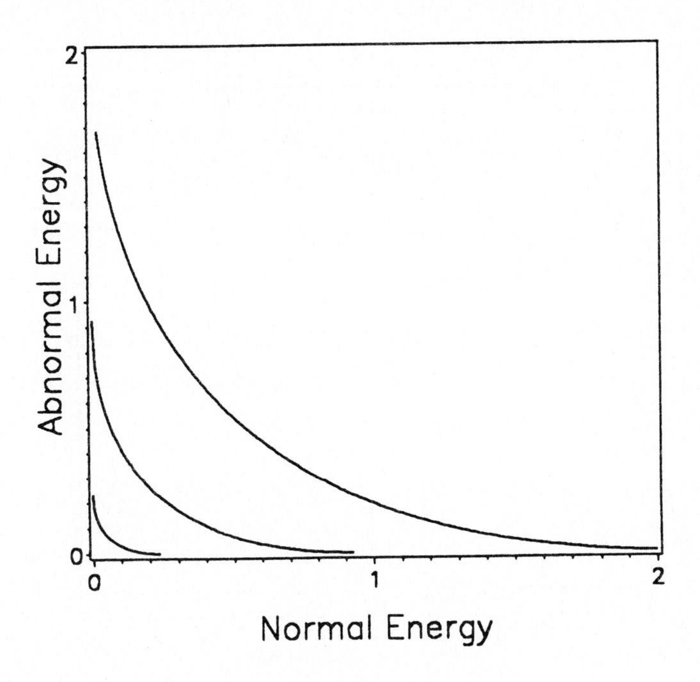

Figure 3.25: Maximum distinguishable energies of excitation for $\bar{g} = 0.5$ (bottom), 1.0 (middle) and 1.5 (top).

This relation is a necessary and sufficient condition on the values of E_1, E_2 and $\bar{g}$ to assure that abnormal transient excitations whose energy of deviation from $\bar{g}$ does not exceed E_2, can always be distinguished from normal performance subject to transient perturbations around zero of energy less than E_1, by a single measurement of the system at time T_f.

For fixed $\bar{g}$, K and T_f, eq. (3.163) expresses a trade-off between the normal and abnormal energies of excitation which can be distinguished. Fig. 3.25 shows the maximum value of E_2 (the energy of abnormal excitation) which can be distinguished from E_1 (the energy of normal excitation). In this figure $K = T_f = 1$, and results for several values of $\bar{g}$ are shown.

The duration of the transient excitation, T_f, also influences which excitation energies are distinguishable. The righthand side of eq. (3.163) is a monotonically increasing function of T_f, which means that the range

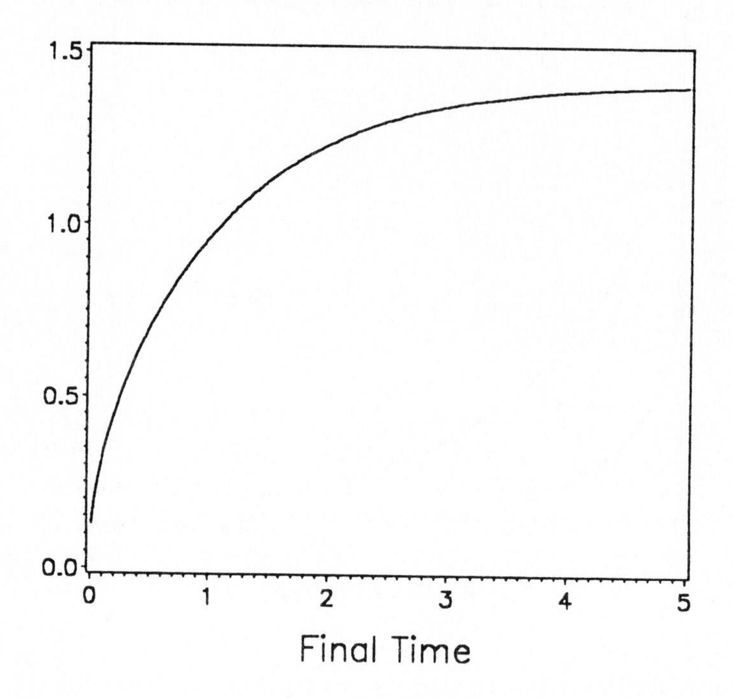

Figure 3.26: $\left[2\left(e^{T_f}-1\right)/\left(e^{T_f}+1\right)\right]^{1/2}$ versus T_f.

of distinguishable excitation energies increases with the duration of the transient. Fig. 3.26 shows the value of the righthand side of eq. (3.163) versus T_f, for $K = \bar{g} = 1$. For large T_f the righthand side of eq. (3.163) asymptotically approaches $\bar{g}\sqrt{2/K}$, as seen in the figure. Also, for very small T_f, the righthand side of eq. (3.163) is proportional to $\bar{g}\sqrt{T_f}$. This indicates that classes of very short transients can be distinguished only if the energies of deviation are small.

Chapter 4

Geometric Imperfections

Images are not arguments, rarely even lead to proof, but the mind craves them, and, of late more than ever, the keenest experimenters find twenty images better than one, especially if contradictory; since the human mind has already learned to deal in contradictions.[1]

4.1 Dynamics of Thin Bars

4.1.1 Introduction

There is a considerable body of literature dealing with the dynamic buckling of bars. In a number of studies a deterministic model of initial imperfections is adopted. Koning and Taub (1934), for example, considered the initial imperfections in the form of a half sine wave, which is the buckling mode of the corresponding static problem. They showed that, if the load is greater than the classical buckling load, the bar deflection grows exponentially in time and remains freely oscillating after unloading, and that the amplitude of the oscillation equals the maximum deflection. If the load is less than the Euler buckling load

[1] Adams, 1918, p 489.

The material in this section was presented at the winter annual meeting of the American Society of Mechanical Engineers, December, 1989, (Ben-Haim and Elishakoff, 1989c).

then the maximum deflection can occur after unloading. Related studies are those of Taub (1934), Gerard and Becker (1952), Meier (1945), Davidson (1953), Hoff (1951), Vol'mir (1966, 1967), Sevin (1960), Hausner and Tso (1962), Huffington (1963), Kaiuk (1965), Lavrent'ev and Ishlinsky (1949), Kornev (1968) and Malyshev (1966). For the general nonlinear theory of dynamic imperfection sensitivity one should consult Budiansky and Hutchinson (1964).

The initial imperfections however are not necessarily proportional to the classical buckling mode, but rather are functions involving uncertainty. Indeed it is unlikely that two bars, even produced by the same manufacturing procedure, will possess the same deviations from the nominally straight state. This uncertainty was studied by Lindberg (1965) using probabilistic analysis. The initial imperfection function was expanded in a Fourier series in terms of the classical buckling modes. The Fourier coefficients were assumed to be normally distributed random variables with zero mean, and with variance proportional to the power spectral density of the initial imperfection. The latter was assumed to be band-limited white-noise. The expectation of the wavelength, measured between the alternate zero crossings, was evaluated (with the assumption of the homogeneity of random imperfections) and wavelength histograms were plotted on the basis of 65 computations.

Elishakoff (1978a,b, 1983) treated the same problem from the perspective of structural reliability. The problem formulation was as follows: Given probabilistic information on the random initial imperfections, find the probabilistic characteristics of the first excursion time, that is, the time required for the bar response to move outside some prescribed safe domain. Closed form results were reported by Elishakoff (1978a, 1983) for the case when the initial imperfection was co-configurational with the buckling mode, whereas for the general case the solution was obtained through the Monte Carlo method. If sufficient probabilistic information is available, then probabilistic analysis enables one to evaluate the reliability of the structure, which is of foremost importance in design. However, in many cases, the probabilistic information on initial imperfections which is needed to determine the reliability is unfortunately unavailable.

The aim of our analysis is to exploit *fragmentary* information which may be available. In particular, we determine the maxima of various parameters characterizing the deflection of a thin bar under dynamic loading conditions, given uncertain initial geometric imperfections. By evaluating the maxima of the dynamic-response parameters as functions of the load and of the uncertainty in the initial imperfection of the bar, it is possible to identify the maximum allowable imperfection for given load. The following parameters are studied.

1. The maximum deflection of the bar as a function of position on the bar and time after onset of the load.

2. The maximum duration for which the deflection exceeds a specified value.

3. The maximum integral deflection along the bar.

4. The maximum integral of the square of the deflection along the bar.

The uncertainty in the initial imperfection profiles will be specified in terms of the variability of the Fourier coefficients of those profiles. The first N Fourier coefficients are assumed to fall in an ellipsoidal set in N-dimensional Euclidean space.

4.1.2 Analytical Formulation

We will adopt the following nomenclature:
x = axial coordinate,
L = beam length,
E = modulus of elasticity,
ρ = density of the bar material,
I = moment of inertia,
A = cross-sectional area,
$\xi = x/L$, normalized position along the bar, $0 \leq \xi \leq 1$,
$\omega_1 = (\pi/L)^2\sqrt{EI/\rho A}$, fundamental frequency of the bar in the absence of axial compression,
$\tau = \omega_1 t$, normalized time after onset of the loading of the thin bar,
$\Delta = \sqrt{I/A}$, radius of gyration of the bar cross-section,

$\bar{y}(x)$ = initial imperfection function,
$y(x)$ = transverse displacement measured from the initial imperfection function,
$\eta = \bar{y}/\Delta$, nondimensional initial imperfection,
$u = y/\Delta$, nondimensional additional displacement,
ν = total normalized deflection $= \eta + u$,
$P_{cl} = \pi^2 EI/L^2$, classical Euler buckling load,
$\alpha = P/P_{cl}$, load ratio.

With these definitions the differential equation for the motion of the bar is (Elishakoff, 1978b)

$$\frac{\partial^4 u}{\partial \xi^4} + \pi^2 \alpha \frac{\partial^2 u}{\partial \xi^2} + \pi^4 \frac{\partial^2 u}{\partial \tau^2} = -\pi^2 \alpha \frac{\partial^2 \eta}{\partial \xi^2} \tag{4.1}$$

The initial and boundary conditions are:

$$u(\xi, \tau) = \frac{\partial^2 u}{\partial \xi^2} = 0 \quad \text{at} \quad \xi = 0 \quad \text{and} \quad \xi = 1 \tag{4.2}$$

$$u(\xi, \tau) = \frac{\partial u}{\partial \tau} = 0 \quad \text{at} \quad \tau = 0 \tag{4.3}$$

The excitation is a compressive step-loading: The constant load P is applied at time $\tau = 0$ and kept indefinitely. We expand the initial imperfection profile in a Fourier sine series as

$$\eta(\xi) = \sum_{n=1}^{\infty} A_n \sin n\pi\xi \tag{4.4}$$

Similarly, we expand the additional deflection as

$$u(\xi, \tau) = \sum_{n=1}^{\infty} G_n(\tau) \sin n\pi\xi \tag{4.5}$$

Following Elishakoff (1978b) one finds that the Fourier coefficients of the additional deflection profile are

$$G_n(\tau) = A_n \psi_n(\tau) \tag{4.6}$$

where ψ_n is the following function:

$$\psi_n(\tau) = \begin{cases} (\cosh(r_n\tau) - 1)/(1 - \beta_n) & \beta_n < 1 \\ \alpha n^2\tau^2/2 & \beta_n = 1 \\ (\cos(r_n\tau) - 1)/(1 - \beta_n) & \beta_n > 1 \end{cases} \quad (4.7)$$

β_n and r_n are defined as

$$\beta_n = \frac{n^2}{\alpha} \qquad r_n = n\sqrt{\mid n^2 - \alpha \mid} \quad (4.8)$$

Finally one obtains the following expression for the total normalized deflection at position ξ and at time τ as

$$\nu(\xi, \tau) = \sum_{n=1}^{\infty} A_n[1 + \psi_n(\tau)] \sin n\pi\xi \quad (4.9)$$

In the subsequent discussion this series will be truncated at $n = N$.

Uncertainty in the initial imperfection profile is represented by allowing the first N Fourier coefficients of the initial deflection, $A_1, \ldots, A_N$, to vary on an ellipsoidal set of values:

$$Z(\theta, W) = \left\{A = (A_1, \ldots, A_N) : \ A^T W A \leq \theta^2\right\} \quad (4.10)$$

where W is an $N \times N$ positive definite real symmetric matrix and θ^2 is a positive number. The shape and size of the ellipsoid are determined by W and θ^2, which are chosen to represent available information concerning the variability of the Fourier coefficients of the initial deflection profile. The set of extreme points of the set $Z(\theta, W)$ is the ellipsoidal shell:

$$C(\theta, W) = \{A = (A_1, \ldots, A_N) : \ A^T W A = \theta^2\} \quad (4.11)$$

4.1.3 Maximum Deflection

The first dynamic response parameter to be examined is the maximum deflection of the bar. The maximum deflection provides a useful and well-recognized criterion for failure: If the instantaneous displacement exceeds a specified threshold, then failure is presumed to result (Hoff,

1951). The maximum on $Z(\theta, W)$ of the total displacement of the bar at time τ and position ξ is

$$\nu_{\max}(\xi,\tau) = \max_{A \in Z(\theta,W)} \nu(\xi,\tau) \tag{4.12}$$

Because $\nu(\xi,\tau)$ is a linear function of the Fourier coefficients, and because $Z(\theta, W)$ is a convex set, the maximum of ν occurs on the set of extreme points of Z. Thus

$$\nu_{\max}(\xi,\tau) = \max_{A \in C(\theta,W)} \nu(\xi,\tau) \tag{4.13}$$

We can obtain an explicit expression for $\nu_{\max}(\xi,\tau)$ by employing the method of Lagrange multipliers. Let us define an N-dimensional vector φ whose nth element is

$$\varphi_n(\xi,\tau) = [1 + \psi_n(\tau)] \sin n\pi\xi \tag{4.14}$$

Thus, truncating ν in eq. (4.9) after the Nth term, one finds

$$\nu(\xi,\tau) = A^T \varphi(\xi,\tau) \tag{4.15}$$

We wish to maximize $A^T\varphi$ subject to the constraint that A falls on the extreme points of Z. That is, $A^T W A = \theta^2$. Let the Hamiltonian be

$$H = A^T\varphi + \lambda(A^T W A - \theta^2) \tag{4.16}$$

A necessary condition for an extremum is

$$0 = \frac{\partial H}{\partial A} = \varphi + 2\lambda W A \tag{4.17}$$

Combining this relation of the constraint, one finds the maximum total displacement of the bar, at position ξ and time τ, to be

$$\nu_{\max}(\xi,\tau) = \theta\sqrt{\varphi(\xi,\tau)^T W^{-1} \varphi(\xi,\tau)} \tag{4.18}$$

Furthermore, the least favorable initial imperfection spectrum, which maximizes the deflection at position ξ and time τ, is:

$$A_{\text{lf}} = \frac{\theta}{\sqrt{\varphi(\xi,\tau)^T W^{-1} \varphi(\xi,\tau)}} W^{-1} \varphi(\xi,\tau) \tag{4.19}$$

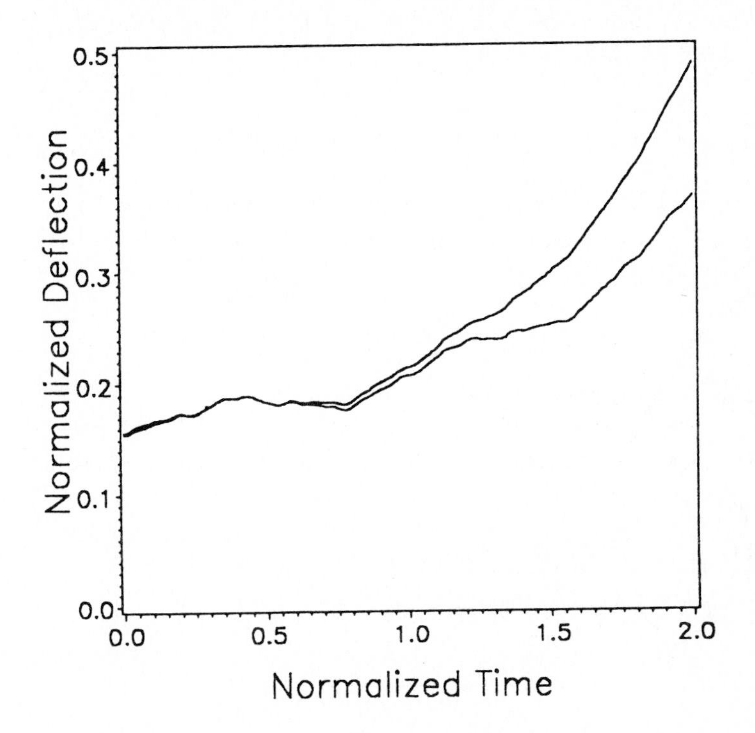

Figure 4.1: Maximum normalized deflection versus normalized time at $\xi = 0.5$ (upper curve) and deflection history reaching maximum deflection at $\tau = 0.5$ (lower curve). Load ratio = 2.

Let us consider a numerical example. The set of allowed initial imperfection spectra is a sphere centered at the origin in 10-dimensional Euclidean space. Thus 10 harmonics are considered and W is the 10×10 identity matrix. The radius of the sphere is $\theta = 0.07$. The upper curve of Fig. 4.1 shows the maximum normalized total deflection as a function of normalized time, at the midpoint of the bar ($\xi = 0.5$). The axial load is twice the classical buckling load, so $\alpha = 2$. Thus, for example, the greatest total deflection which can be achieved at $(\xi, \tau) = (0.5, 2.0)$ is 0.49, or nearly half the radius of gyration of the bar. The lower curve shows the deflection history of a bar which reaches maximum deflection at $\tau = 0.5$. This bar deflects to the maximum value of $\nu = 0.18$ at $\tau = 0.5$ and deflects less than the limiting value for the remainder of

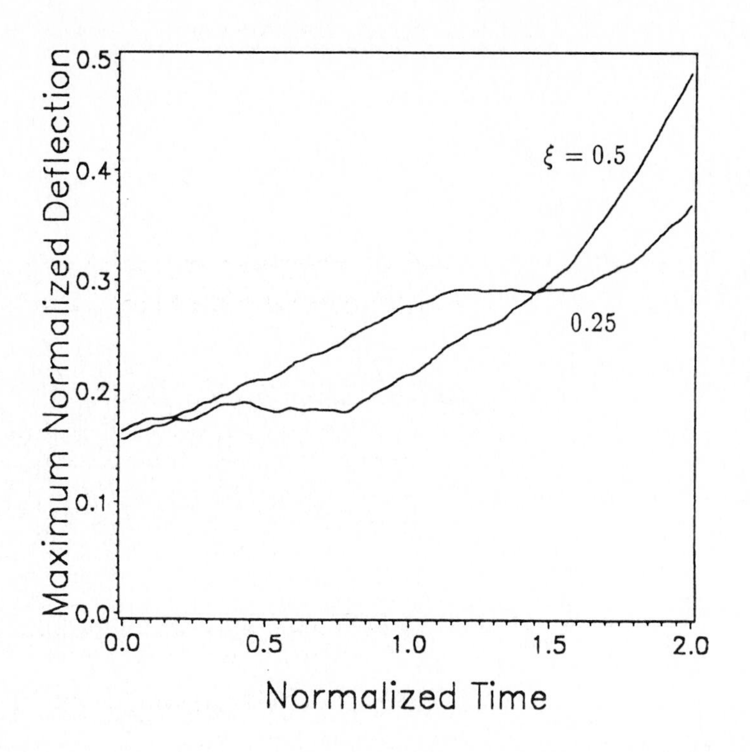

Figure 4.2: Maximum normalized deflection versus normalized time at $\xi = 0.5$ and $\xi = 0.25$. Load ratio $= 2$.

the history.

Fig. 4.2 shows the maximum total deflection at two points on the bar ($\xi = 0.25$ and $\xi = 0.5$) as a function of time. The axial load ratio is $\alpha = 2$. The deflection at $\xi = 0.25$ exceeds the deflection at $\xi = 0.5$ up to about $\tau = 1.5$. At later times the maximum possible midpoint deflection overcomes the deflection at one-fourth the length. Diagrams of this kind allow one to find the envelope of maximum displacement as functions of both space and time.

Fig. 4.3 shows the maximum normalized deflection at the midpoint of the bar as a function of time, for several values of the axial load ratio. The greatest possible deflection for loads less than the classical buckling load are small. The maximum deflection increases sharply as the axial load is increased. The conclusion here is that the least

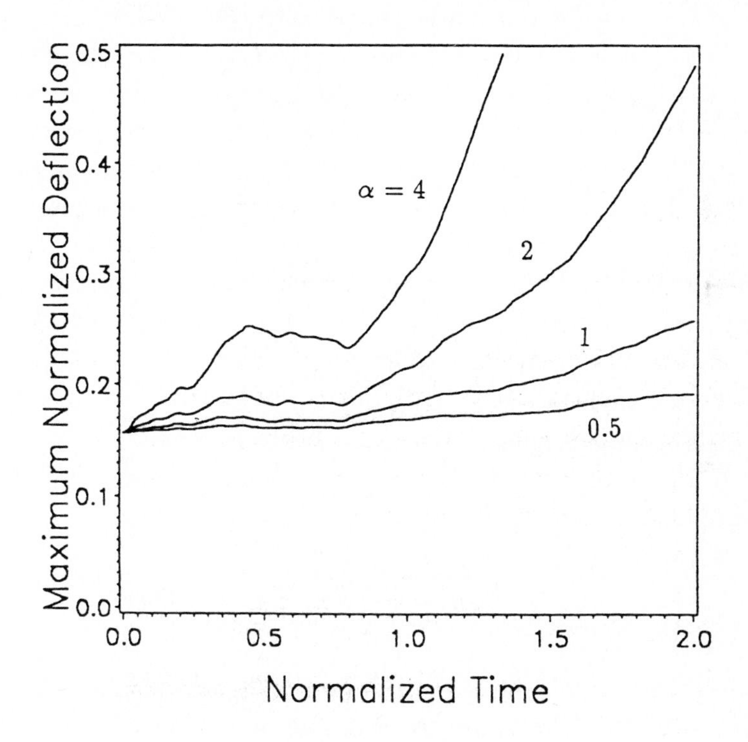

Figure 4.3: Maximum normalized deflection versus normalized time at $\xi = 0.5$ for various values of the load ratio.

favorable displacement can be evaluated for any specified value of the load.

4.1.4 Duration Above a Threshold

The maximum deflection of the bar provides a useful and well recognized criterion for failure. However, the duration for which the displacement is maintained may also be a relevant parameter for characterizing the phenomenon of failure under a dynamic load. For example, one can imagine that failure results from a displacement of extended duration, even if that displacement falls short of the threshold value.

We will evaluate the maximum duration of displacement above the level μ by determining whether or not displacement above μ can be sustained for intervals of successively greater duration. We wish to know if an initial displacement spectrum exists such that the displacement at point ξ exceeds the value μ during the time interval $[\tau_1, \tau_2]$. By varying τ_1 and τ_2, we will find the greatest duration of displacement exceeding μ. Choose K instants T_k such that $\tau_1 \leq T_1 < \cdots < T_K \leq \tau_2$. Let $\varphi^k = \varphi(\xi, T_k)$, $k = 1, \ldots, K$, where the elements of the N-dimensional vector φ are defined in eq. (4.14). We wish to know if there exists an allowed initial displacement spectrum A such that the displacement at position ξ equals or exceeds μ at instants $T_1, \ldots, T_K$. By choosing K sufficiently large with respect to the highest frequency component in the displacement spectrum, we are able to determine whether or not the displacement remains at or above μ during $[\tau_1, \tau_2]$.

Stated mathematically, we wish to know if a real solution, A, exists for the following set of inequalities:

$$A^T W A \leq \theta^2 \tag{4.20}$$

$$A^T \varphi^k \geq \mu, \quad k = 1, \ldots, K \tag{4.21}$$

The first inequality states that A belongs to the set $Z(\theta, W)$ of initial imperfection spectra. The latter K relations assert that the displacement equals or exceeds μ at the instants $T_1, \ldots, T_K$.

Relations (4.20) and (4.21) define a compact set. We wish to determine whether or not this set is empty. Let ζ be an arbitrary N-dimensional vector. Thus $\zeta^T A$ is a continuous real function on this set. A continuous real function achieves a minimum and a maximum value on a compact set. Consequently, if a vector A exists which satisfies (4.20) and (4.21), then a vector exists which satisfies (4.20) and (4.21) and which extremizes $\zeta^T A$. So the existence of a vector satisfying relations (4.20) and (4.21) can be determined by seeking the extremum of $\zeta^T A$ subject to the constraints expressed by (4.20) and (4.21). If no extremum exists, then the set defined by relations (4.20) and (4.21) is empty.

Constraints (4.20) and (4.21) can be expressed as equalities by introducing additional real variables, $h_0, \ldots, h_K$:

$$A^T W A + h_0^2 = \theta^2 \tag{4.22}$$

$$A^T \varphi^k - h_k^2 = \mu \quad , \quad k = 1, \ldots, K \tag{4.23}$$

To find the extrema of $\zeta^T A$ subject to these constraints, define the Hamiltonian:

$$H = \zeta^T A + \lambda_0 \left(A^T W A + h_0^2 - \theta^2\right) + \sum_{k=1}^{K} \lambda_k (A^T \varphi^k - h_k^2 - \mu) \quad (4.24)$$

Necessary conditions for an extremum are

$$0 = \frac{\partial H}{\partial A} = \zeta + 2\lambda_0 W A + \sum_{k=1}^{K} \lambda_k \varphi^k \quad (4.25)$$

$$0 = \frac{\partial H}{\partial h_0} = 2\lambda_0 h_0 \quad (4.26)$$

$$0 = \frac{\partial H}{\partial h_k} = -2\lambda_k h_k \quad , \quad k = 1, \ldots, K \quad (4.27)$$

Eqs. (4.22), (4.23) and (4.25) – (4.27) define $N+2K+2$ relations for the $N+2K+2$ unknowns: $A, h_0, h_1, \ldots, h_K, \lambda_0, \lambda_1, \ldots, \lambda_K$. The solution of these equations is facilitated by the fact that the extrema of a linear function on a convex set occur on the boundaries of the set. Eqs. (4.26) and (4.27) imply that $\lambda_k = 0$ if $h_k \neq 0$. For the problem in question, the solution is usually on only a single boundary.

4.1.5 Maximum Integral Displacements

We have discussed two criteria for failure: maximum displacement and duration of displacement beyond a specified threshold. Other dynamical parameters which can be readily calculated may prove useful in characterizing the phenomenon of dynamic failure. We will consider two integral parameters. Integral parameters are simpler than the position-dependent parameters considered earlier, since they represent global or average response.

The integral (span-averaged) displacement along the bar at time τ is

$$\delta_1(\tau) = \int_0^1 \nu(\xi, \tau) d\xi \quad (4.28)$$

$$= \frac{2}{\pi} \sum_{n=0}^{\infty} \frac{A_{2n+1}}{2n+1} \left[1 + \psi_{2n+1}(\tau)\right] \quad (4.29)$$

Let Ξ_1 be a vector of dimension N whose even-indexed terms are zero and whose odd-indexed terms are $2[1+\psi_n(\tau)]/(n\pi)$, for $n = 1,3,5,\ldots$. Then $\delta_1 = A^T\Xi_1$. To determine the maximum integral deflection we must maximize δ_1 subject to the requirement that A belongs to $Z(\theta, W)$. The extrema of the linear function δ_1 occur on the boundary of the convex set Z. Thus the constraint that A belongs to Z can be expressed as

$$A^T W A = \theta^2 \tag{4.30}$$

To maximize δ_1, define the Hamiltonian as

$$H = A^T\Xi_1 + \lambda\left(A^T W A - \theta^2\right) \tag{4.31}$$

Necessary conditions for an extremum of δ_1 are

$$0 = \frac{\partial H}{\partial A} = \Xi_1 + 2\lambda W A \tag{4.32}$$

Combining this with eq. (4.30) yields the following expression for λ :

$$\lambda = \pm\frac{1}{2\theta}\sqrt{\Xi_1^T W^{-1}\Xi_1} \tag{4.33}$$

Now multiply eq. (4.32) on the left by A^T to obtain

$$A^T\Xi_1 = -2\lambda A^T W A \tag{4.34}$$

The lefthand side is δ_1, while eq. (4.30) implies that $A^T W A = \theta^2$ on the right. Combining eqs. (4.33) and (4.34) yields the following expression for the maximum of δ_1 :

$$\delta_{1,\max}(\tau) = \theta\sqrt{\Xi_1^T W^{-1}\Xi_1} \tag{4.35}$$

Fig. 4.4 illustrates the maximum span-averaged displacement versus the normalized time, for various values of the load ratio, α. The number of harmonics is 10 (so the series in eq. (4.29) is truncated after $n = 10$.) Also W is the identity matrix and $\theta = 0.07$. Figs. 4.3 and 4.4 are generally similar, except that the latter figure is smoother due to the fact that $\delta_{1,\max}$ is a span-averaged rather than a local quantity. This suggests that not too much information is lost in considering the integral, span-averaged displacement rather than the position-dependent

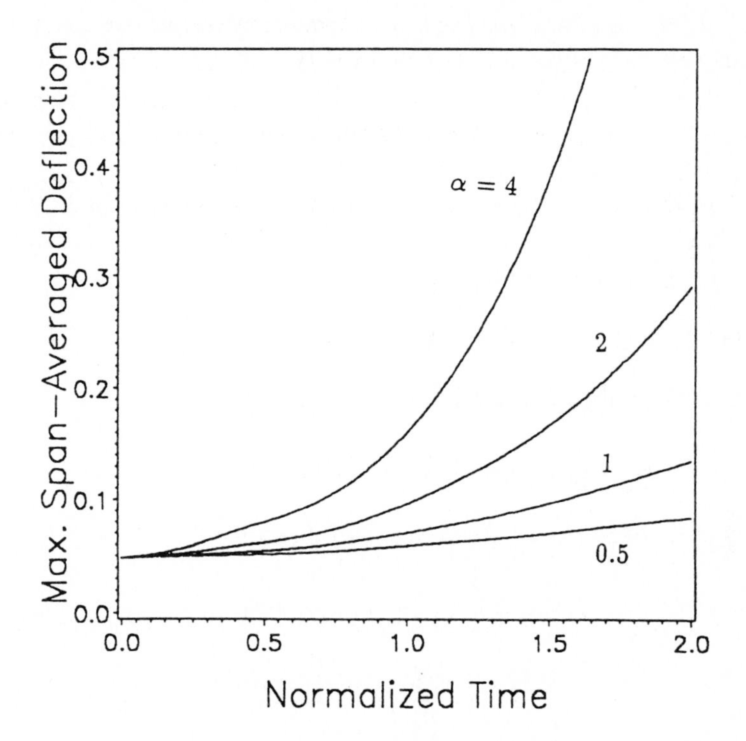

Figure 4.4: Maximum integral (span-averaged) deflection versus normalized time for various values of the load ratio.

displacement. The span-averaged displacement seems to be a useful and concise dynamic failure criterion.

Let us now calculate the integral square displacement along the bar at time τ :

$$\delta_2(\tau) = \int_0^1 \nu^2(\xi,\tau)d\xi \tag{4.36}$$

$$= \frac{1}{2}\sum_{n=1}^{\infty} A_n^2[1+\psi_n(\tau)]^2 \tag{4.37}$$

We wish to find the maximum of $\delta_2(\tau)$ as A varies on the set of allowed initial imperfection profiles $Z(\theta, W)$. Let Ξ_2 be a diagonal matrix whose nth diagonal element is $[1+\psi_n(\tau)]^2$. Examination of eq. (4.7) shows that these diagonal elements are always positive, so Ξ_2 is a positive definite matrix. Thus δ_2 can be represented as:

$$\delta_2 = \frac{1}{2} A^T \Xi_2 A \tag{4.38}$$

We seek to maximize δ_2 subject to the requirement that A belongs to $Z(\theta, W)$. This constraint can be expressed as an equality by including an additional unknown variable h :

$$A^T W A + h^2 = \theta^2 \tag{4.39}$$

To maximize δ_2, define the Hamiltonian as

$$H = \frac{1}{2} A^T \Xi_2 A + \lambda \left(A^T W A + h^2 - \theta^2 \right) \tag{4.40}$$

Necessary conditions for an extremum of δ_2 are

$$0 = \frac{\partial H}{\partial A} = \Xi_2 A + 2\lambda W A \tag{4.41}$$

$$0 = \frac{\partial H}{\partial h} = 2\lambda h \tag{4.42}$$

If the extremum occurs in the interior of $Z(\theta, W)$ then $h > 0$ and eq. (4.42) implies that $\lambda = 0$. Since Ξ_2 is a non-singular matrix, eq. (4.41) then implies that $A = 0$, hence $\delta_2 = 0$. Since Ξ_2 is a positive definite matrix this is the minimum of δ_2 rather than the maximum. Thus the maximum of δ_2 occurs on the boundary of Z, so $A^T W A = \theta^2$. Since $A \neq 0$, eq. (4.41) implies that $\Xi_2 + 2\lambda W$ must be a singular matrix. Thus the determinant of the latter matrix vanishes:

$$|\Xi_2 + 2\lambda W| = 0 \tag{4.43}$$

Since W is positive definite this relation can be written as

$$\left| W^{-1} \right| \cdot \left| W^{-1} \Xi_2 + 2\lambda I \right| = 0 \tag{4.44}$$

Hence the extrema of δ_2 occur on the boundary of Z when -2λ is a characteristic root of the matrix $W^{-1}\Xi_2$. To obtain an expression for the maximum of δ_2 multiply eq. (4.41) by A^T on the left to obtain

$$\frac{1}{2} A^T \Xi_2 A = -\lambda A^T W A \tag{4.45}$$

The lefthand side is δ_2, while $A^TWA = \theta^2$ on the righthand side and λ satisfies eq. (4.44).

As a specific example, let W be diagonal, with diagonal elements $1/w_n^2$, $n = 1, \ldots, N$. Then δ_2 attains an extremal value when λ assumes any of the following N values:

$$\lambda = -\frac{1}{2} w_n^2 [1 + \psi_n(\tau)]^2 \quad , \quad n = 1, \ldots, N \tag{4.46}$$

Employing eq. (4.45) one finds the global maximum of δ_2 to be

$$\delta_{2,\max}(\tau) = \frac{\theta^2}{2} \max_{n=1,\ldots,N} w_n^2 [1 + \psi_n(\tau)]^2 \tag{4.47}$$

Fig. 4.5 illustrates the square root of the maximum integral square displacement versus the normalized time, for various values of the load ratio, α. The number of harmonics is chosen to be 10 (so the series in eq. (4.37) is truncated after $n = 10$.) Also W is the identity matrix and $\theta = 0.07$. These assumptions, both the diagonality of W or number and indices of the dominant harmonics, could be easily eliminated if sufficient experimental information is available for the description of the uncertainty of the bar imperfections.

We have studied the deflection of a thin elastic bar subject to uncertain initial geometric imperfections and step loading. The usual approach to this analysis is based on knowledge of the probability distribution of the initial geometric imperfections. In the absence of such knowledge, a probabilistic approach can not be employed. Moreover, the reliability function resulting from a probabilistic analysis turns out to be highly sensitive to the parameters of the probability distribution employed. Consequently, if one arbitrarily adopts a probability distribution, the resulting reliability function is likely to be inaccurate.

The initial deflection Fourier coefficients were postulated to fall inside an N-dimensional ellipsoid. Two position-dependent dynamic parameters were studied: maximum deflection and maximum duration of deflection. Failure is presumed to result when the deflecton exceeds an empirically determined threshold. However, extended deflection below the threshold can also result in failure. We have shown that these parameters are readily evaluated without relying on knowledge of the

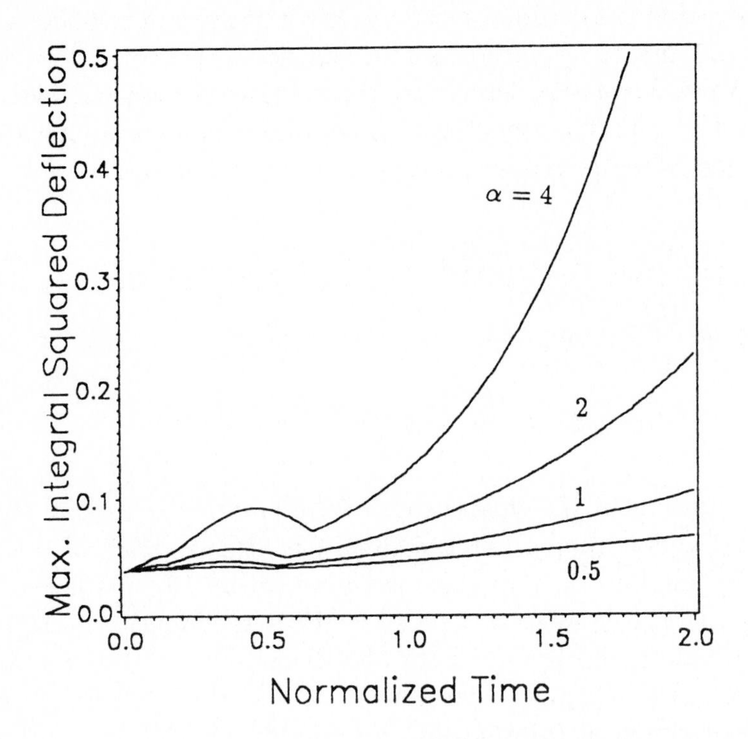

Figure 4.5: Maximum integral squared deflection versus normalized time for various values of the load ratio.

probability distribution of the initial imperfection: information which is often lacking. Two integral dynamic parameters for concise representation of the response have also been studied: the maximum integral, span-averaged displacement and the maximum integral squared displacement. These integral parameters are position-independent and thus simpler to standardize than the position-dependent parameters. Our results suggest that, from the point of view of the general trend, the integral parameters are nearly as representative of the bar deflection as the position-dependent deflection, and thus provide a concise criterion for failure. The main conclusion is that both pointwise or integral characteristics can be readily evaluated.

To sum up, the convex models of uncertainty employed in this section for the impact problem have been shown to be useful analytical tools when probabilistic measures of uncertainty are unavailable. These models by no means replace the probabilistic approach. The latter should be used when sufficient data are available for probabilistic description of the uncertain quantities. However if this data is unavailable for any reason, then convex modelling becomes an alternative.

4.2 Impact Loading of Thin Shells

4.2.1 Introduction

Dynamics and failure of isotropic, unstiffened cylindrical shells are dealt with in a number of studies. These papers could be classified according to various criteria. One of them is whether the model of the initial imperfections and/or elastic properties is deterministic or probabilistic. Deterministic models of imperfections were adopted by Coppa and Nash (1964), Bieneck *et al* (1966), Anderson and Lindberg (1968), Kornev (1969, 1972), Kornev and Solodovnikov (1972) and by Zincik and Tennyson (1980). Probabilistic models of uncertainty were utilized by Lindberg and Herbert (1966) who dealt with random initial imperfections, and by Astil *et al* (1972) who studied concrete cylinders with random ultimate strength and density, via the Monte Carlo method. This method was also utilized by Elishakoff (1978, 1983) for evaluating failure due to the impact of bars with random imperfections, but with otherwise deterministic properties. For an extensive bibliography one should consult the recent monograph by Lindberg and Florence (1987).

Another distinctive mark of the available literature is the variety of failure criteria applied. Kornev (1972) suggests to replace a bar under impact by a single-degree-of-freedom system, described by a maximum characteristic exponent, so that the displacements will rise exponentially in time. This vibration mode is different from the fundamental one. For shells (Kornev and Markin (1974)) the maximum characteristic exponent corresponds to the multiplicity of motions. In very thin shells, they find a condensation point with an infinite set of motions

The material of this section was presented at the Euromech Symposium 250, Como, Italy, July, 1989, (Elishakoff and Ben-Haim, 1989).

with almost coincident characteristic exponents. In this case, the shell can not be replaced with a single-degree-of-freedom system. Rather, multi-degree-of-freedom analysis should be used. A "practical" criterion for the safe operation of the structure under dynamic loading was suggested by Hoff (1965): a safe operation region of the structure is one where "...admissible finite disturbances of its initial state of static or dynamic equilibrium are followed by displacements whose magnitude remains within allowable bounds during the required lifetime of the structure". Applying Hoff's definition, one may assume that a shell with initial imperfections fails under axial forces when its dynamic response (deflection, strain or stress) first reaches an upper-bound level Q^+ or a lower-bound level, $-Q^-$, where Q^+ and Q^- are prescribed positive numbers which represent borderlines between failure and safe operation. However, the rigorous determination of values for Q^+ and Q^- remains an unresolved challenge.

We will study impact loading of isotropic shells with initial imperfections, but we will not adopt any specific failure criterion. To simulate an often encountered practical situation, we will assume that only limited information is available for the initial imperfections of the shells. In particular, we will assume that the initial imperfection Fourier coefficients fall within a given ellipsoidal set, which could be determined experimentally. With this limited information we determine the bounds of the total response. The ratio of the failure boundary to the worst-case response will define the minimum factor of safety, depending upon the definition of the failure.

4.2.2 Basic Equations

The differential equation governing the axisymmetric motion of a thin cylindrical shell reads

$$D\frac{\partial^4 w}{\partial x^4} + N\frac{\partial^2 w}{\partial x^2} + \rho h\frac{\partial^2 w}{\partial t^2} + \frac{Eh}{R^2}w = -N\frac{\partial^2 w_0}{\partial x^2} \tag{4.48}$$

where: $w_0(x)$ is the initial imperfection function, $w(x,t)$ is the additional shell displacement, x the axial coordinate, t the time, $D = \frac{Eh^3}{12(1-\nu^2)}$ is the flexural stiffness, E is Young's modulus, h is the shell thickness, R the shell radius, ν is Poisson's ratio, N the axial loading, and ρ the shell material density. The shell is simply supported at its ends. Eq. (4.48) is supplemented by the following boundary and initial conditions:

$$w(x,t) = \frac{\partial^2 w}{\partial x^2} = 0 \quad \text{at} \quad x = 0 \quad \text{and} \quad x = L \tag{4.49}$$

$$w(x,t) = \frac{\partial w}{\partial t} = 0 \quad \text{at} \quad t = 0 \tag{4.50}$$

We introduce the following nondimensional quantities:

$$\xi = \frac{x}{L} \ , \ \tau = \omega_1 t \ , \ \alpha = \frac{N}{N_{cl}} \tag{4.51}$$

where ξ is a non-dimensional axial coordinate, L is the length, ω_1 is the first natural frequency of the shell in vacuo and N_{cl} is the classical buckling load of the shell:

$$\omega_1^2 = \frac{D\pi^4}{\rho h L^4} + \frac{E}{\rho R^2} \ , \ N_{cl} = \frac{D}{L^2}\gamma(\beta) \tag{4.52}$$

where

$$\beta = \frac{EhL^4}{DR^2} \ , \ \gamma(\beta) = k^2(\beta)\pi^2 + \frac{\beta}{k^2(\beta)\pi^2} \tag{4.53}$$

$k(\beta)$ is an integer denoting the number of half sine-waves in the axial direction. The value of γ which determines the classical buckling load in eq. (4.52) is the minimum with respect to k for fixed β. For $\beta \gg 1$, k can be treated as a continuous parameter and $\gamma(\beta)$ is minimized at

$$k^4 = \frac{\beta}{\pi^4} \tag{4.54}$$

yielding the minimum of γ as:

$$\gamma^*(\beta) = \frac{4L^2}{Rh}\sqrt{3(1-\nu^2)} \tag{4.55}$$

and

$$N_{cl} = \frac{Eh^2}{R\sqrt{3(1-\nu^2)}} \tag{4.56}$$

which is a well known expression for the classical buckling load of a thin shell. The differential equation (4.48) becomes

$$\frac{\partial^4 u}{\partial \xi^4} + \alpha\gamma^*(\beta)\frac{\partial^2 u}{\partial \xi^2} + (\beta + \pi^4)\frac{\partial^2 u}{\partial \tau^2} + \beta u = -\alpha\gamma^*(\beta)\frac{\partial^2 u_0}{\partial \xi^2} \tag{4.57}$$

where

$$u = \frac{w}{h} \; , \; u_0 = \frac{w_0}{h} \tag{4.58}$$

are nondimensional additional displacement and initial imperfections, respectively.

For the simply supported shell we expand the initial imperfection profile in a Fourier sine series

$$u_0(\xi) = \sum_{n=1}^{\infty} A_n \sin n\pi\xi \tag{4.59}$$

Similarly, we expand the additional displacement of the shell in a series as

$$u(\xi, \tau) = \sum_{n=1}^{\infty} G_n(\tau) \sin n\pi\xi \tag{4.60}$$

which leads to the following differential equation for $G_n(\tau)$:

$$\frac{d^2 G_n}{d\tau^2} + \frac{n^2\pi^2}{\beta + \pi^4}\left[n^2\pi^2 - \alpha\gamma^* + \frac{\beta}{n^2\pi^2}\right] G_n(\tau) = \frac{\alpha\gamma^*}{\beta + \pi^4} n^2\pi^2 A_n \tag{4.61}$$

The particular solution reads

$$G_n = \begin{cases} \dfrac{\alpha\gamma^* A_n}{n^2\pi^2 - \alpha\gamma^* + \beta/n^2\pi^2} & , \quad \alpha\gamma * \neq n^2\pi^2 + \dfrac{\beta}{n^2\pi^2} \\ \dfrac{\alpha\gamma^* \tau^2 n^2\pi^2 A_n}{2(\beta + \pi^4)} & , \quad \alpha\gamma^* = n^2\pi^2 + \dfrac{\beta}{n^2\pi^2} \end{cases} \tag{4.62}$$

To obtain the homogeneous solution of eq. (4.61) we note that the characteristic equation is obtained by putting $G_n = Ce^{r\tau}$, yielding:

$$r^2 + \frac{n^2\pi^2}{\beta + \pi^4}\left[n^2\pi^2 - \alpha\gamma^* + \frac{\beta}{n^2\pi^2}\right] = 0 \tag{4.63}$$

The homogeneous solution has an oscillatory character if $\alpha\gamma* < n^2\pi^2 + \beta/n^2\pi^2$; it is exponentially increasing in time if $\alpha\gamma^* > n^2\pi^2 + \beta/n^2\pi^2$; and is linearly increasing in time if $\alpha\gamma^* = n^2\pi^2 + \beta/n^2\pi^2$.

The differential equation (4.61) is supplemented by the initial conditions

$$G_n(0) = \frac{dG_n(0)}{d\tau} = 0 \tag{4.64}$$

so that $G_n(\tau)$ can be represented as

$$G_n(\tau) = A_n\psi_n(\tau) \tag{4.65}$$

where

$$\psi_n(\tau) = \frac{\alpha\gamma^* A_n}{\alpha\gamma^* - n^2\pi^2 - \beta/n^2\pi^2}[\cosh(r_n\tau) - 1] \quad \text{for } \alpha\gamma^* > n^2\pi^2 + \frac{\beta}{n^2\pi^2} \tag{4.66}$$

$$\psi_n(\tau) = \frac{\alpha\gamma^* A_n}{\alpha\gamma^* - n^2\pi^2 - \beta/n^2\pi^2}[\cos(r_n\tau) - 1] \quad \text{for } \alpha\gamma^* < n^2\pi^2 + \frac{\beta}{n^2\pi^2} \tag{4.67}$$

$$\psi_n(\tau) = \frac{\alpha\gamma^* n^2\pi^2\tau^2 A_n}{2(\beta + \pi^2)} \quad \text{for } \alpha\gamma^* = n^2\pi^2 + \frac{\beta}{n^2\pi^2} \tag{4.68}$$

where

$$r_n = \frac{n\pi}{\sqrt{\beta + \pi^4}}\sqrt{\left|n^2\pi^2 - \alpha\gamma^* + \frac{\beta}{n^2\pi^2}\right|} \tag{4.69}$$

The total normalized displacement $v(\xi, \tau)$ of the shell is

$$v(\xi,\tau) = u_0(\xi) + u(\xi,\tau) = \sum_{n=1}^{\infty} A_n[1 + \psi_n(\tau)] \sin n\pi\xi \tag{4.70}$$

4.2.3 Extremal Displacement

Let us assume that we have only limited information for characterizing the initial imperfections. In particular, the only information is that the dominant N initial imperfection Fourier coefficients in eq. (4.59) fall within an ellipsoidal set:

$$Z(\Omega,\theta)=\left\{A^T=(A_{m_1},A_{m_2},\ldots,A_{m_N}):\ A^T\Omega A\leq\theta^2\right\} \tag{4.71}$$

where Ω is a positive definite symmetric matrix, θ^2 is a positive constant and $m_1, m_2, \ldots, m_N$ are the indices of the dominant imperfection amplitudes. We will assume that the rest of the Fourier coefficients vanish identically. Thus eq. (4.70) becomes:

$$v(\xi,\tau)=\varphi(\xi,\tau)^T A \tag{4.72}$$

Here φ is an N-vector whose nth element is

$$\varphi_n(\xi,\tau)=[1+\psi_{m_n}(\tau)]\sin m_n\pi\xi \tag{4.73}$$

The problem is formulated as follows: given an imperfection ellipsoid of the initial imperfections, find the initial imperfection vector which maximizes the total displacement. We will denote this maximizing vector A_{worst}. This maximum is formally represented as:

$$\hat{v}(\xi,\tau)=\max_{A\in Z(\Omega,\theta)} v(\xi,\tau) \tag{4.74}$$

The set of extreme points of $Z(\Omega,\theta)$ is the ellipsoidal shell

$$C(\Omega,\theta)=\{A:\ A^T\Omega A=\theta^2\} \tag{4.75}$$

The set $Z(\Omega,\theta)$ is a convex set and $v(\xi,\tau)$ is a linear function of A. Hence the maximum deflection will be reached on the extreme points of $Z(\Omega,\theta)$, i.e. on the ellipsoidal shell $C(\Omega,\theta)$. In other words:

$$\hat{v}(\xi,\tau)=\max_{A\in C(\Omega,\theta)} v(\xi,\tau) \tag{4.76}$$

A closed form expression for $\hat{v}(\xi,\tau)$ is obtained by the method of Lagrange multipliers. Define the Hamiltonian as:

$$H(A)=\varphi^T A+\lambda(A^T\Omega A-\theta^2) \tag{4.77}$$

For the extremum we require

$$\frac{\partial H}{\partial A}=0=\varphi+2\lambda\Omega A \tag{4.78}$$

or

$$A = -\frac{1}{2\lambda}\Omega^{-1}\varphi \tag{4.79}$$

Combining this expression with the constraint on A: $A^T\Omega A = \theta^2$, one finds:

$$A^T\Omega A = \frac{1}{4\lambda^2}\varphi^T\Omega^{-1}\varphi = \theta^2 \tag{4.80}$$

or

$$\lambda^2 = \frac{1}{4\theta^2}\varphi^T\Omega^{-1}\varphi \tag{4.81}$$

This yields the vector causing the worst (maximum) deflection:

$$A_{\text{worst}} = \frac{\theta}{\sqrt{\varphi^T\Omega^{-1}\varphi}}\Omega^{-1}\varphi \tag{4.82}$$

Combining eqs. (4.72) and (4.82) one obtains the maximum displacement as:

$$\hat{v}(\xi,\tau) = \varphi^T A_{\text{worst}} = \theta\sqrt{\varphi(\xi,\tau)^T\Omega^{-1}\varphi(\xi,\tau)} \tag{4.83}$$

It is remarkable that A_{worst} is a function of the vector $\varphi(\xi,\tau)$, which means that A_{worst} depends on time τ and on the space coordinate ξ. At first glance this appears to be a surprising conclusion because the vector A represents the initial imperfections which are "built in" to the structure prior to use and do not vary in time. The dependence of A_{worst} on ξ and τ should be interpreted as follows: The initial imperfection vector which maximizes the total displacement at (ξ,τ) depends on ξ and τ. In other words, *different* initial imperfections will maximize the total displacements at the *different* ξ and τ.

4.2.4 Numerical Example

Let us consider a two-mode imperfection model for which the deflections in the mid-region of the shell tend to be magnified:

$$u_0(\xi) = A_m \sin m\pi\xi + A_{m+4}\sin(m+4)\pi\xi \tag{4.84}$$

where m is a positive integer and the vector $A^T = (A_m, A_{m+4})$ falls within the ellipsoidal set $Z(\Omega,\theta)$ with

$$\Omega = \begin{bmatrix} 1 & 1/2 \\ 1/2 & 1/3 \end{bmatrix} \tag{4.85}$$

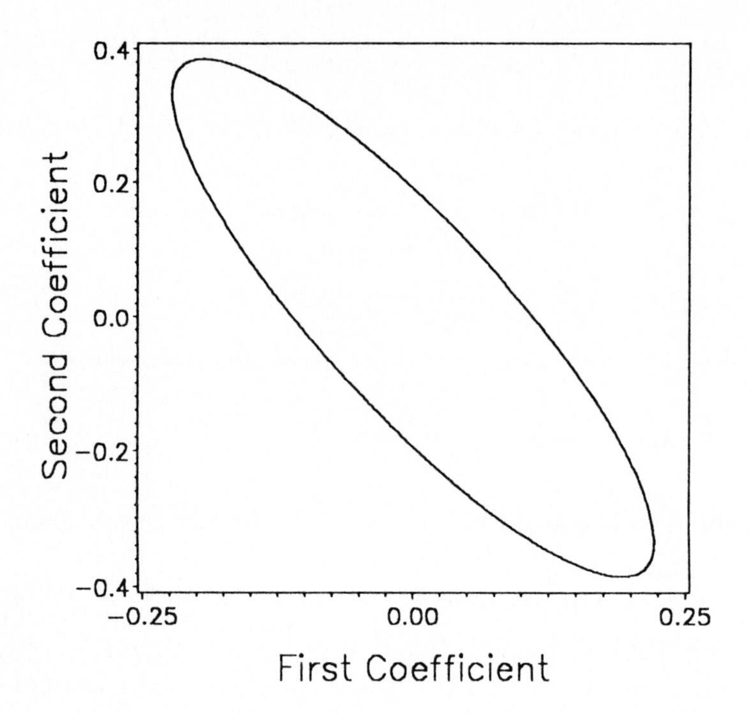

Figure 4.6: Ellipsoidal set of Fourier coefficients for a two-mode imperfection.

and $\theta = 1/9$. The allowed ellipsoidal set for the initial imperfections is then

$$Z(\Omega,\theta) = \left\{ (A_m, A_{m+4}) : A_m^2 + A_m A_{m+4} + \frac{1}{3} A_{m+4}^2 \leq \frac{1}{81} \right\} \tag{4.86}$$

The boundary and interior of the ellipse in Fig. 4.6 define the region of allowed values of (A_m, A_m+4). The extreme points of $Z(\Omega,\theta)$ constitute an ellipse whose elements satisfy:

$$A_m^2 + A_m A_{m+4} + \frac{1}{3} A_{m+4}^2 = \frac{1}{81} \tag{4.87}$$

Equation (4.81) becomes

$$\lambda^2 = \frac{81}{4} \varphi^T \Omega^{-1} \varphi \tag{4.88}$$

$$= \frac{81}{4}\{\varphi_m \ \varphi_{m+4}\}\begin{bmatrix} 4 & -6 \\ -6 & 12 \end{bmatrix}\begin{Bmatrix} \varphi_m \\ \varphi_{m+4} \end{Bmatrix} \tag{4.89}$$

$$= \frac{81}{4}(4\varphi_m^2 - 12\varphi_m\varphi_{m+4} + 12\varphi_{m+4}^2) \tag{4.90}$$

From eq. (4.82) one finds the maximizing coefficients to be:

$$A_{m,\text{worst}} = \frac{1}{9}\frac{1}{\sqrt{4\varphi_m^2 - 12\varphi_m\varphi_{m+4} + 12\varphi_{m+4}^2}}(4\varphi_m - 6\varphi_{m+4}) \tag{4.91}$$

$$A_{m+4,\text{worst}} = \frac{1}{9}\frac{1}{\sqrt{4\varphi_m^2 - 12\varphi_m\varphi_{m+4} + 12\varphi_{m+4}^2}}(-6\varphi_m + 12\varphi_{m+4}) \tag{4.92}$$

The maximum deflection is:

$$\hat{v}(\xi, \tau) = \frac{1}{9}\sqrt{4\varphi_m^2 - 12\varphi_m\varphi_{m+4} + 12\varphi_{m+4}^2} \tag{4.93}$$

We fix the geometric characteristics of the shell at

$$\frac{L}{R} = 3 \quad , \quad \frac{R}{h} = 250 \tag{4.94}$$

so that

$$\beta = 5.52825 \times 10^7 \quad , \quad \gamma^* = 14870.44048 \quad , \quad k = \left[\frac{\beta^{1/4}}{\pi}\right] = 27 \tag{4.95}$$

where $[x]$ denotes the greatest integer not exceeding x.

To get some insight we will consider three subcases. The load ratio throughout these subcases will be fixed at $\alpha = 2$.

Subcase 1. Let us concentrate first on the case when the initial imperfection spectrum does not contain the number of half-waves of the shell during the static, classical buckling. That is, neither m nor $m + 4$ equals k. Specifically we take $m = 1$, so that we have nonzero Fourier coefficients $A_{1,\text{worst}}$ and $A_{5,\text{worst}}$ of the "excited" modes. We will be interested in the displacement of the shell in the middle cross section $\xi = 1/2$, with

$$u_0(1/2) = A_1 + A_5 \tag{4.96}$$

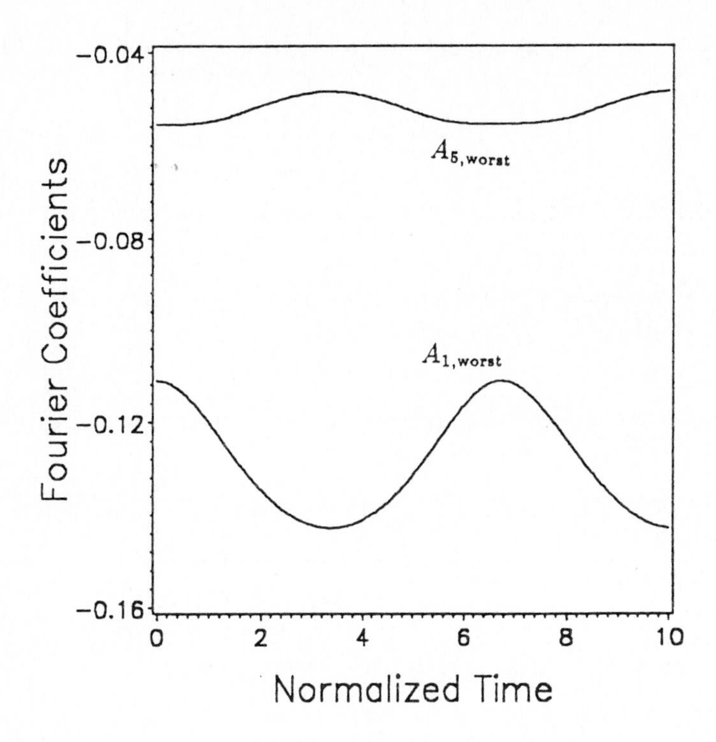

Figure 4.7: Fourier coefficients, $A_{1,\mathrm{worst}}$ and $A_{5,\mathrm{worst}}$, which maximize the deflection of the shell in subcase 1.

and

$$\varphi_1 = 1 + \psi_1(\tau) \quad , \quad \varphi_5 = 1 + \psi_5(\tau) \tag{4.97}$$

Now

$$\psi_1 = 0.00533798(1 - \cos 0.99734\tau) \tag{4.98}$$

$$\psi_5 = 0.1528643(1 - \cos 0.93186\tau) \tag{4.99}$$

Fig. 4.7 depicts $A_{1,\mathrm{worst}}$ and $A_{5,\mathrm{worst}}$, the elements of the vector which maximizes the displacement. The corresponding maximum displacement $\hat{\nu}(1/2, \tau)$ is displayed in Fig. 4.8. The maximum displacement is oscillatory in time because ψ_1 and ψ_2 are harmonic functions.

Subcase 2. Here one of the initial imperfection components is co-configurational to the classical buckling mode, namely $m = 27$ and

$$u_0(1/2) = -A_{27} - A_{31} \tag{4.100}$$

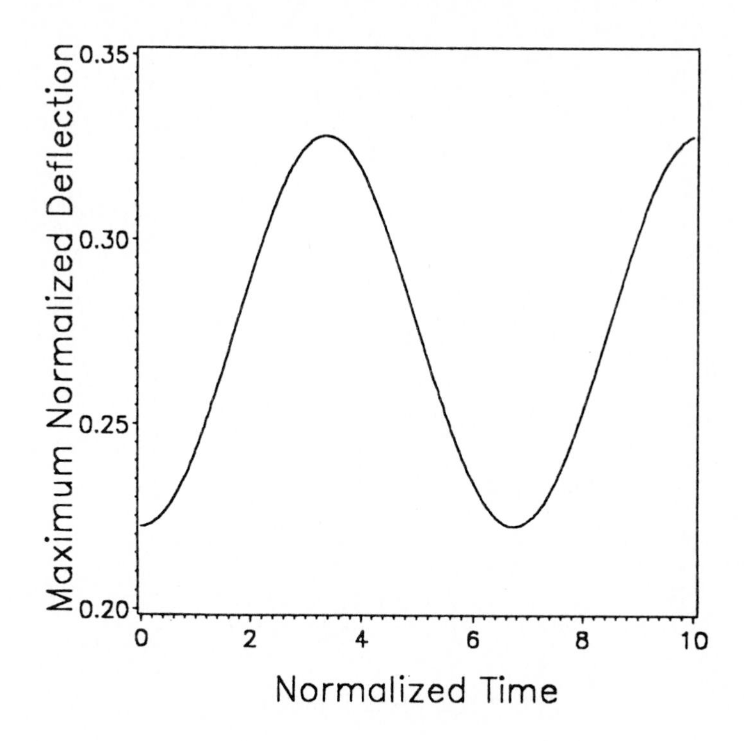

Figure 4.8: Maximum deflection of the shell for subcase 1.

and

$$\varphi_{27} = 1 + \psi_{27}(\tau) \quad , \quad \varphi_{31} = 1 + \psi_{31}(\tau) \tag{4.101}$$

In these circumstances

$$\psi_{27} = 2.00110798(\cosh 1.390798\tau - 1) \tag{4.102}$$

$$\psi_{31} = 2.0613898(\cosh 1.57330963\tau - 1) \tag{4.103}$$

Fig. 4.9 portrays the Fourier coefficients, $A_{27,\mathrm{worst}}$ and $A_{31,\mathrm{worst}}$, while Fig. 4.10 is associated with the maximum displacement $\hat{\nu}(1/2,\tau)$. In this case the Fourier coefficients which maximize the response have more pronounced variation than for the case when the first and fifth modes are dominant in the imperfection profile. In these circumstances the worst response rapidly becomes unbounded with increase in time.

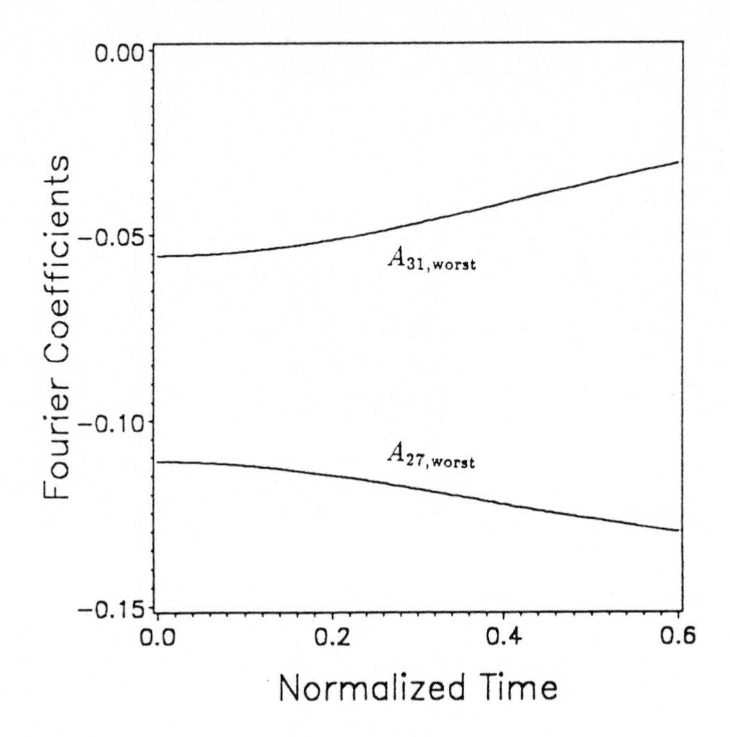

Figure 4.9: Fourier coefficients, $A_{27,\text{worst}}$ and $A_{31,\text{worst}}$, which maximize the deflection of the shell in subcase 2.

Subcase 3. Here we have $m = 13$, with

$$u_0(1/2) = A_{13} + A_{17} \tag{4.104}$$

The appropriate ψ functions are

$$\psi_{13} = 5.86512(1 - \cos 0.391145\tau) \tag{4.105}$$

$$\psi_{17} = 3.9617823(\cosh 0.622352\tau - 1) \tag{4.106}$$

Fig. 4.11 shows $A_{13,\text{worst}}$ and $A_{17,\text{worst}}$ as functions of time. The maximum displacement, $\hat{\nu}(1/2, \tau)$, is given in Fig. 4.12. Here too the worst response tends to infinity with increase in time, since ψ_{17} increases exponentially.

This section presents the convex-modelling approach to describing the axisymmetric dynamics of a geometrically imperfect thin cylindrical

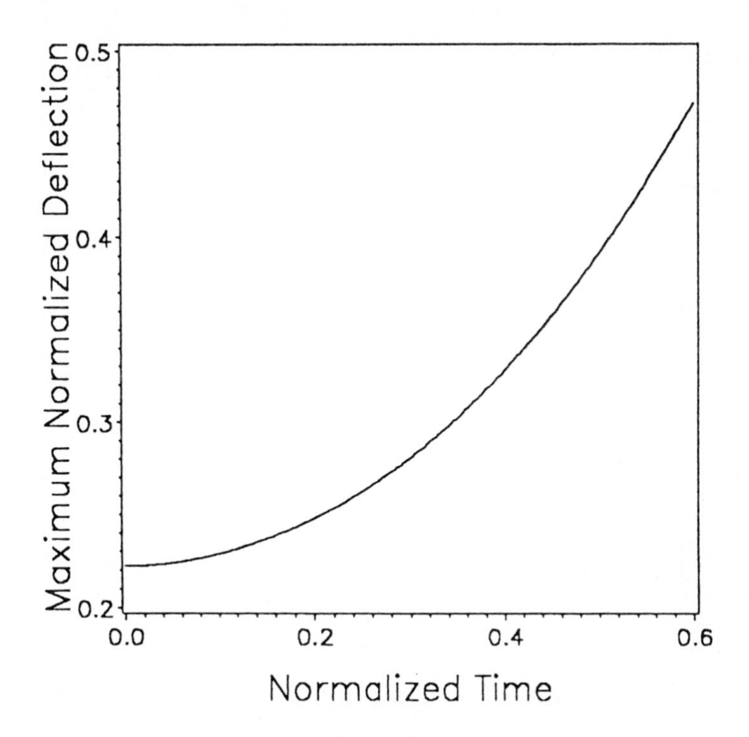

Figure 4.10: Maximum deflection of the shell for subcase 2.

shell under impact loading. The initial geometrical imperfection profile of the shell is represented by a Fourier series. The uncertainty in the imperfection profile is represented by specifying an ellipsoidal set which bounds the variation of the dominant Fourier coefficients. Expressions are derived for the maximum deflection of the shell and for the vector of Fourier coefficients which maximizes the deflection.

If one adopts Hoff's definition of failure then one immediately finds the following factor of safety:

$$fs = \frac{Q^+}{\hat{v}(\xi, \tau)} \tag{4.107}$$

where Q^+ is the maximum allowable deflection of the shell and $\hat{v}$ is the greatest possible deflection for any imperfection consistent with information used in defining the set of initial imperfection profiles. Even

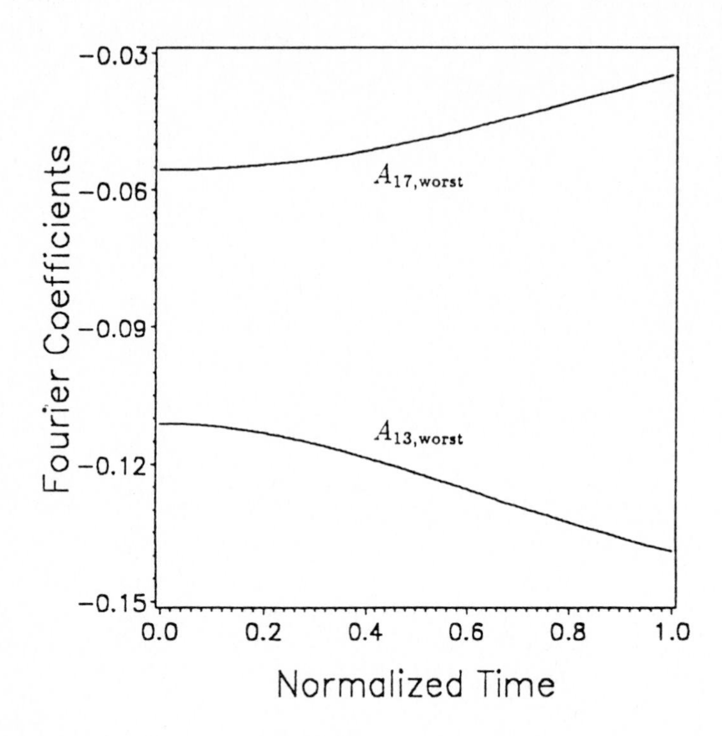

Figure 4.11: Fourier coefficients, $A_{13,\text{worst}}$ and $A_{17,\text{worst}}$, which maximize the deflection of the shell in subcase 3.

if Hoff's criterion of failure is not used, the maximum deflection, $\hat{v}$, can be used to characterize the performance of the structure subject to incompletely specified initial imperfections.

4.3 Buckling of Thin Shells

4.3.1 Introduction

Thin-walled shells, which are among the most utilized structural elements in engineering, possess the unfortunate property of being sensitive to initial geometrical imperfections (Koiter, 1945; Budiansky and Hutchinson, 1979). This means that the buckling load may be significantly reduced due to deviations of the real shell from its ideal, nominal

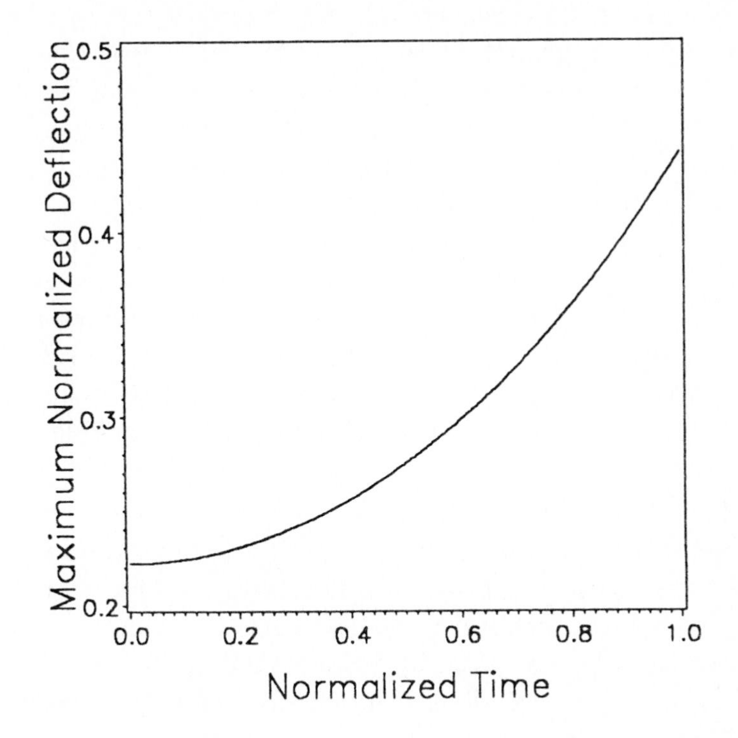

Figure 4.12: Maximum deflection of the shell for subcase 3.

counterpart. The experimental buckling load sometimes amounts to no more than one-tenth of the buckling load of a perfect shell. Existing computer codes are able to predict the buckling load of a structure with specified initial imperfections. However, due to the very nature of the manufacturing process, it is hard to imagine that two identical shells are ever produced, even by the same manufacturing process. Consequently is it imperative to consider the effect on the buckling load of uncertainty in the imperfections of the shell.

Engineers are accustomed to using empirical "knockdown factors"

This section represents an extended version of our paper, Ben-Haim and Elishakoff, 1989a. Different portions of this material were presented at the XVIIth International Congress of Theoretical and Applied Mechanics, Grenoble, France, August, 1988 and at the International Symposium on Stochastic Structural Dynamics, in honor of Y.K. Lin, University of Illinois, Urbana, October, 1988.

(NASA, 1979) in order to accomodate the large discrepancy between theoretical and experimental values of the buckling load. The knockdown factor, when multiplied by the classical buckling load for the perfect structure, yields an estimated lower bound of the buckling load for the imperfect structure. Knockdown factors are often adopted as the lower bound of the buckling loads obtained experimentally for a range of distinct structures, materials and manufacturing processes. Such an approach has several drawbacks. It would seem that the estimates should constantly be updated to include new experimental results. Furthermore, this approach mixes shells produced by rough manufacturing procedures (and therefore associated with a greater reduction of the buckling load) with those produced by more refined techniques (and hence buckling at greater load). This implies that the design of shells with low initial imperfections will be overly conservative.

A natural way to deal with uncertainty in the initial imperfections is to employ a stochastic approach. Apparently the first probabilistic analysis of initial imperfections, treated as random variables with specified joint distribution, was given by Bolotin (1958). The next step was treatment by Fraser and Budiansky (1969) of initial imperfections as random fields with given mean and autocorrelation function. These approaches have been bridged by Elishakoff (1979, 1980) in the context of the Monte Carlo method. For acquaintance with pertinent results one may consult the review paper by Amazigo (1976) and the monographs by Roorda (1980) and Elishakoff (1983).

Probabilistic analysis treats the initial imperfections as random functions of the space coordinates y, z of the shell. Let $\eta(y, z)$ represent the deviation of the shell from its nominal shape at point y, z. If one knows an analytic relation between the buckling load Λ and the initial imperfection function $\eta(y, z)$

$$\Lambda = \Psi(\eta(y, z)) \tag{4.108}$$

then one can relate the probabilistic characteristics of Λ with those of η, resulting in an expression for the probability density of the buckling load. Except for the simplest cases, there is no analytical relation of type (4.108) available in the literature. Usually, the initial imperfection function is expanded in a Fourier series:

$$\eta(y,z) = \sum_{i,j} A_{ij}\varphi_{ij}(y,z) \tag{4.109}$$

where A_{ij} are the Fourier coefficients and φ_{ij} is a complete set of functions. Then available computer codes (Arbocz and Babcock, 1974) yield relations of the type:

$$\Lambda = \Psi(A_{ij}) \tag{4.110}$$

For probabilistic methods of dealing with eq. (4.110), with appropriate bibliography, one can consult Elishakoff and Arbocz (1985) and Elishakoff *et al* (1987).

The aim of our analysis is to exploit fragmentary information (which is usually all that is available) about the initial imperfection of thin shells, in order to determine the buckling loads which may be expected. Explicitly, the minimum buckling load will be determined as a function of parameters which characterize the range of possible initial imperfection profiles of the shell. Non-probabilistic convex models of uncertainty in the initial imperfections will be employed. This means that an infinite set of initial profiles will be adopted on the basis of available data, and then the minimum of the buckling load on this set will be sought. A significant result of this analysis is a theoretical estimate of the knockdown factor. In addition, the knockdown factor will be expressed as a function of simple manufacturing specifications. It will be seen that the set-theoretic approach to modelling uncertainty in the initial imperfections of the shell is quite flexible and allows one to examine imperfection sensitivity from various perspectives.

The range of variation of the initial imperfection profiles will be modelled in several distinct ways. In sections 4.3.2 and 4.3.3 the uncertainty in the initial imperfection profiles will be quantified in terms of the variability of the Fourier coefficients of those profiles. The N most significant Fourier coefficients are assumed to fall in an ellipsoidal set in N-dimensional Euclidean space. The minimum buckling load is then evaluated as a function of the shape of the ellipsoid. In section 4.3.4 the uncertainty in the initial imperfection profile is expressed as a uniform bound on the deviation of the surface from its nominal value. Thus the initial imperfection profiles are integrable functions which satisfy a uniform bound. The minimum buckling load will be determined as a function of the uniform bound on the initial imperfection. This uniform bound on the initial imperfections can be viewed as a shape tolerance, so that the buckling load is related to a manufacturing specification of the shell. In section 4.3.5 the initial imperfection profiles are allowed to vary within an envelope. This enables the designer to study the buckling sensitivity of the shell as a function of requiring the manufacture of the shell to adhere to different tolerances in different areas of the shell. The ideas developed here are used to evaluate the knockdown factor in section 4.3.6. Finally, a comparison of first- and second-order analyses is presented in section 4.3.7.

Following Elishakoff *et al* (1987), the change of the buckling load with the initial imperfection profile will be studied as a perturbation problem. Both first- and second-order variations from a mean or typical initial imperfection will be considered. The buckling load of the mean initial imperfection will be evaluated on the basis of a numerical nonlinear buckling code.

4.3.2 Bounded Fourier Coefficients: First-Order Analysis

Let x be a vector whose components are the N dominant Fourier coefficients of the initial imperfection profile of a thin shell. Furthermore, let $\Psi(x)$ represent the buckling load for a shell whose initial imperfection profile has x as its Fourier spectrum. Let x^o be a nominal Fourier imperfection spectrum. For example, x^o may correspond to the average imperfection spectrum. Shells which have been manufactured and handled under similar conditions have experienced forces which are likely to produce patterns of distortion common to all the shells. The common features of the simulated shell imperfections shown in Fig. 2.1 are quite evident. Consequently, the average imperfection spectrum is unlikely to be zero.

The buckling limit for an initial imperfection spectrum $x^o + \zeta$, to first order in ζ, is:

$$\Psi(x^o + \zeta) = \Psi(x^o) + \sum_{n=1}^{N} \frac{\partial \Psi(x^o)}{\partial x_n} \zeta_n \tag{4.111}$$

We will evaluate the lower limit of the buckling load as ζ varies on an ellipsoidal set of initial imperfection spectra. For convenience of notation let us define:

$$\varphi^T = \left(\frac{\partial \Psi(x^o)}{\partial x_1}, \ldots, \frac{\partial \Psi(x^o)}{\partial x_N} \right) \tag{4.112}$$

where the superscript T means matrix transposition.

The deviation ζ from the nominal initial imperfection spectrum is assumed to vary on the following ellipsoidal set:

$$Z(\alpha, \omega) = \left\{ \zeta : \sum_{n=1}^{N} \frac{\zeta_n^2}{\omega_n^2} \leq \alpha^2 \right\} \tag{4.113}$$

where the size parameter α and the semiaxes $\omega_1, \ldots, \omega_N$ are based on experimental data, obtainable from initial imperfection data banks. Thus $Z(\alpha, \omega)$ can be chosen to represent a realistic ensemble of shells. The lowest buckling load which can be obtained from any of the shells in this ensemble is expressed formally as the minimum of expression (4.111) on the set Z :

$$\mu(\alpha, \omega) = \min_{\zeta \in Z(\alpha, \omega)} \left(\Psi(x^o) + \varphi^T \zeta \right) \tag{4.114}$$

$\mu(\alpha, \omega)$ is the buckling load of the "weakest" shell in the ensemble Z which is constructed to represent a realistic range of shells. It will be recognized from the discussion in section 4.3.1 that the ratio of μ to the classical buckling load will correspond to the empirical knockdown factor. This will be discussed further in section 4.3.6.

Eq. (4.114) calls for finding the minimum of the linear functional $\varphi^T \zeta$ on the convex set $Z(\alpha, \omega)$. This extreme value will occur on the set of extreme points of Z, which is the collection of vectors $c = (c_1, \ldots, c_N)$ in the following set:

$$C(\alpha, \omega) = \left\{ c : \sum_{n=1}^{N} \frac{c_n^2}{\omega_n^2} = \alpha^2 \right\} \tag{4.115}$$

Thus the minimum buckling load, eq. (4.114), becomes:

$$\mu(\alpha,\omega) = \min_{c\in C(\alpha,\omega)} \left(\Psi(x^o) + \varphi^T c\right) \tag{4.116}$$

Define Ω as an $N \times N$ diagonal matrix whose nth diagonal element is $1/\omega_n^2$. Then, as seen from eq. (4.115), we must minimize $\varphi^T c$ subject to the constraint:

$$f(c) \equiv c^T \Omega c - \alpha^2 = 0 \tag{4.117}$$

We will proceed by the method of Lagrangian multipliers (Bryson and Ho, 1975). Define the Hamiltonian as:

$$H(c) = \varphi^T c + \gamma f(c) \tag{4.118}$$

where γ is a constant multiplier whose value must be determined, and φ is defined in eq. (4.112). For an extremum we require that the derivative of the Hamiltonian vanish:

$$0 = \frac{\partial H}{\partial c} = \varphi + 2\gamma\Omega c \tag{4.119}$$

which implies:

$$c = -\frac{1}{2\gamma}\Omega^{-1}\varphi \tag{4.120}$$

Substituting this into the constraint, eq. (4.117), yields the following expression for the multiplier:

$$\gamma^2 = \frac{1}{4\alpha^2}\varphi^T\Omega^{-1}\varphi \tag{4.121}$$

from which we find that the extremal values of the vector c are:

$$c = \pm\frac{\alpha}{\sqrt{\varphi^T\Omega^{-1}\varphi}}\Omega^{-1}\varphi \tag{4.122}$$

We now find that the minimum buckling load, given in eq. (4.116), becomes:

$$\mu(\alpha,\omega) = \Psi(x^o) - \alpha\sqrt{\varphi^T\Omega^{-1}\varphi} \tag{4.123}$$

It is significant that this analysis yields an explicit relationship between the minimum buckling load and the characteristics of the initial

Fourier Coef.	B-1	B-2	B-3	B-4	ω_n	$\partial\Psi/\partial x_i$
a_2	−0.010809	−0.027238	0.089906	−0.017560	0.0315	0.09668
a_4	0.022578	−0.007836	−0.025508	−0.009239	0.0174	0.00340
$b_{1,2}$	0.417400	0.392870	0.741280	0.222900	0.1870	−0.01854
$b_{1,6}$	−0.077872	−0.143490	0.174830	0.077668	0.0853	−0.05687
$b_{1,8}$	−0.263690	−0.009405	0.112470	0.101510	0.1520	−0.24686
$b_{1,10}$	0.036568	0.043628	−0.245610	−0.008853	0.1190	−0.08183
$b_{2,3}$	−0.101290	0.034018	−0.064766	−0.001887	0.0526	−0.01314
$b_{2,11}$	0.009732	−0.008685	−0.028261	0.013545	0.0166	−0.07173

Table 4.1: Shell Data from Elishakoff *et al* (1987).

imperfections, as represented in the parameters α and $\omega_1, \ldots, \omega_N$. Since Ω is a diagonal matrix eq. (4.123) can be written explicitly as:

$$\mu(\alpha, \omega) = \Psi(x^o) - \alpha\sqrt{\sum_{n=1}^{N}\left(\omega_n \frac{\partial\Psi(x^o)}{\partial x_n}\right)^2} \qquad (4.124)$$

From this relation one recognizes that significant reduction in the buckling load results from large sensitivity of the nominal buckling load to Fourier coefficients whose semiaxes in the imperfection ellipsoid are large. Furthermore one notices that the minimum buckling load depends linearly on the overall size, α, of the imperfection ellipsoid, and non-linearly on its shape, $\omega_1, \ldots, \omega_N$ and on the partial derivatives $\partial\Psi(x^o)/\partial x_n$.

Let us consider the numerical evaluation of eq. (4.124), based on partial derivatives of Ψ with respect to 8 significant Fourier coefficients (from Elishakoff *et al* 1987). The buckling load with the nominal imperfection profile is 0.746 (in units of the classical buckling load) for the shells in question. The diagonal elements of the matrix Ω are chosen as the mean squared deviations from the average of the corresponding Fourier coefficients of the four B-shells studied by Elishakoff *et al* (1987). These measured values of the Fourier coefficients for these four shells, the values of $\omega_1, \ldots, \omega_8$ and the derivatives of Ψ are presented here in Table 4.1.

The linear variation of the minimum buckling load with the size

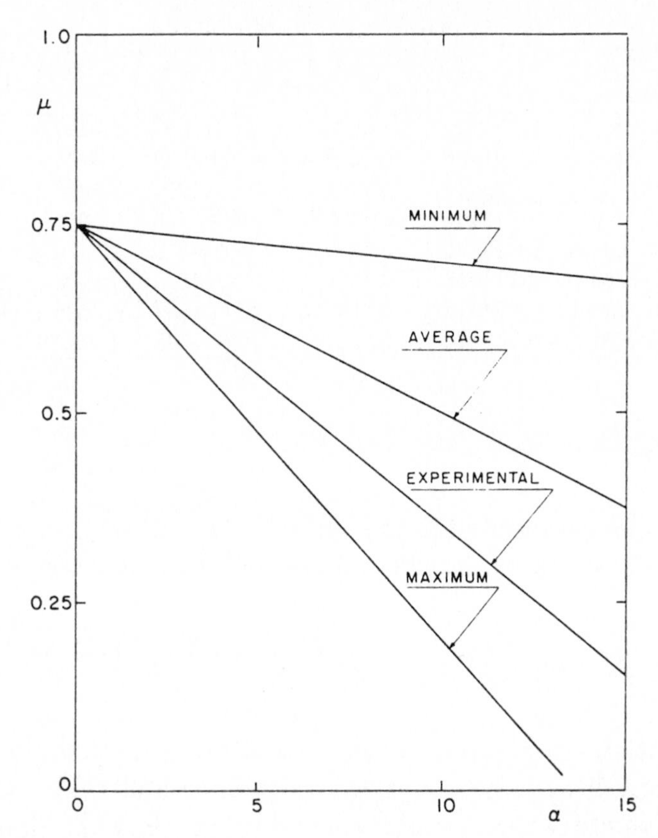

Figure 4.13: Minimum buckling load, μ, as a function of the size, α, of the imperfection ellipsoid, for four different sets of semiaxes.

parameter α is shown in fig. 4.13. The four lines are based on different choices of the semiaxes of the imperfection ellipsoid. The curve labeled "experimental" employs the empirical values of $\omega_1, \ldots, \omega_8$ in Table 4.1; for the other three curves the imperfection sets are spheroids whose radii equal the minimum, the maximum and the average of the semi-axes in Table 4.1. Eq. (4.124) indicates that the minimum buckling load is sensitive to the size and shape of the imperfection ellipsoid. To demonstrate this we evaluate μ as the imperfection ellipsoid varies from the ellipsoid of Table 4.1 to a spheroid of equal volume. The radius of the equivalent spheroid is $\omega_s = 0.0585$. The semi-axes vary parametrically with a control parameter p as $p\omega_n + (1-p)\omega_s$ for $n = 1, \ldots, 8$. The minimum buckling load, μ, is displayed in fig. 4.14 versus the parameter

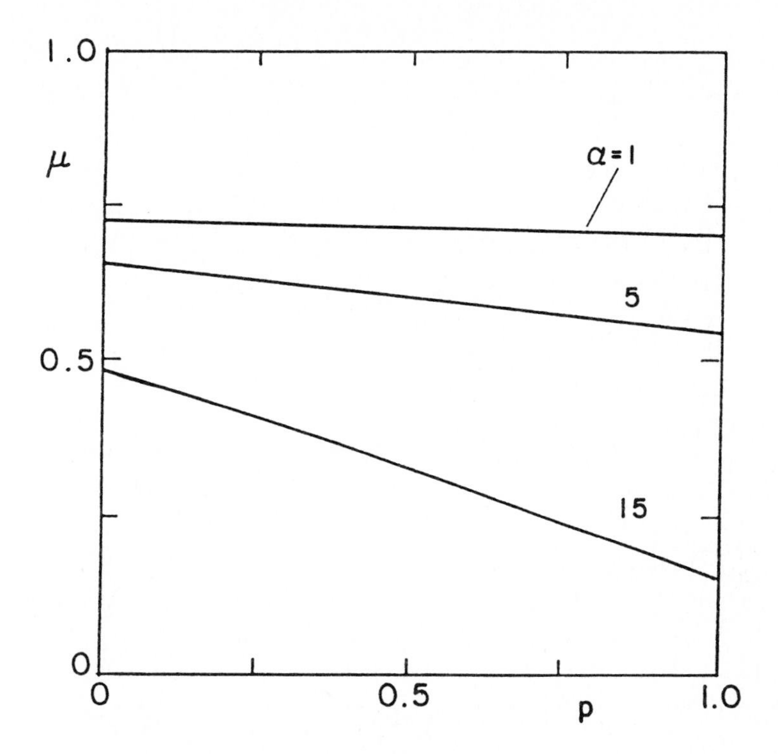

Figure 4.14: Minimum buckling load, μ, as a function of the shape of the imperfection ellipsoid, for three values of α.

p, for three values of the ellipsoid size, α.

Eq. (4.124) indicates that the minimum buckling load is sensitive to the values of the derivatives of Ψ as well as to the size and shape of the imperfection ellipsoid. In fig. 4.15 we evaluate μ as the imperfection ellipsoid varies from the ellipsoid of Table 4.1 to a spheroid of equal volume, and as the derivatives of Ψ vary from the values in Table 4.1 to the average of those derivatives. The radius of the equivalent spheroid is 0.0585 as before, and the average of the eight derivatives of Ψ listed in Table 4.1 is $\bar{d} = -0.04861$. The semi-axes vary parametrically with the parameter p as in fig. 4.14, while the derivatives vary parametrically as $p(\partial\Psi/\partial x_n) + (1-p)\bar{d}$ for $n = 1, \ldots, 8$. The minimum buckling load, μ, is displayed in fig. 4.15 versus the parameter p, for three values of the ellipsoid size, α.

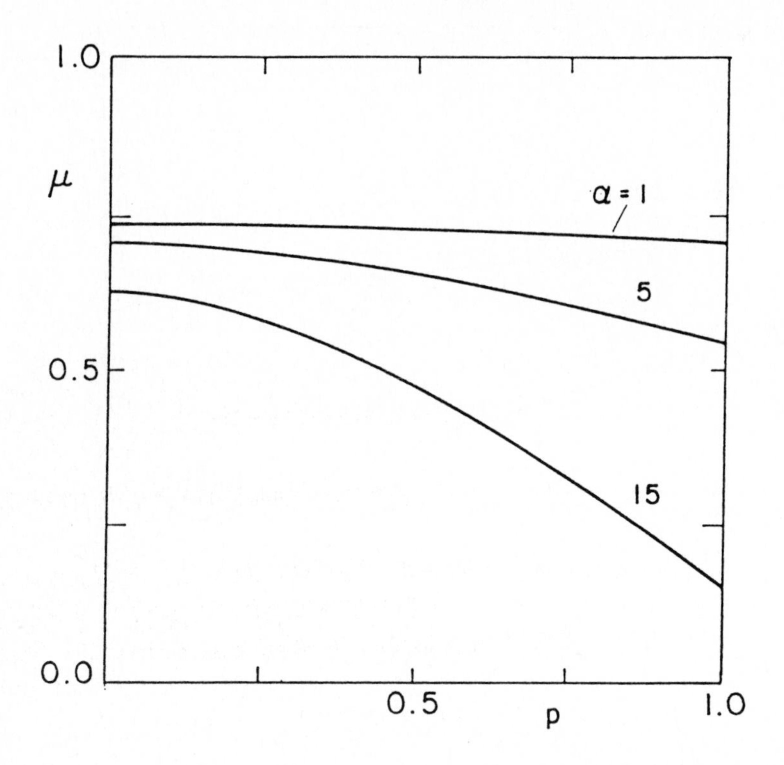

Figure 4.15: Minimum buckling load, μ, as a function of the shape of the imperfection ellipsoid and of the derivatives of Ψ, for three values of α.

It should be noted that, although eqs. (4.123) and (4.124) are written in closed form, $\Psi(x^o)$ as well as $\partial\Psi(x^o)/\partial x_n$ must be determined numerically from existing non-linear computer codes. Thus the latest numerical sophistications can be directly incorporated in this analysis.

4.3.3 Bounded Fourier Coefficients: Second-Order Analysis

In the previous section a first-order expansion of the non-linear buckling load was studied. In this section a second order expansion is considered. Let $Z(\alpha, \omega)$ and φ be defined as in section 4.3.2 and define the $N \times N$

matrix Ξ whose elements are:

$$\xi_{mn} = \frac{\partial^2 \Psi(x^o)}{\partial x_m \partial x_n} \tag{4.125}$$

Let x^o be a nominal initial imperfection spectrum. The buckling limit for an initial imperfection $x^o + \zeta$, to second order in ζ, is:

$$\Psi(x^o + \zeta) = \Psi(x^o) + \varphi^T \zeta + \frac{1}{2} \zeta^T \Xi \zeta \tag{4.126}$$

As before, we are interested in the minimum buckling load on the ensemble Z of shells. This minimum may be expressed formally as:

$$\mu(\alpha, \omega) = \min_{\zeta \in Z(\alpha,\omega)} \left(\Psi(x^o) + \varphi^T \zeta + \frac{1}{2} \zeta^T \Xi \zeta \right) \tag{4.127}$$

This is a non-linear optimization problem with an inequality constraint. Define Ω as in section 4.3.2. The inequality constraint on the deviations ζ from the nominal initial imperfection profile, embodied in the set $Z(\alpha, \omega)$, becomes:

$$f(\zeta) \equiv \zeta^T \Omega \zeta - \alpha^2 \le 0 \tag{4.128}$$

Again the method of Lagrangian multipliers will be adopted to solve this problem. Under the new circumstances the Hamiltonian will be defined as:

$$H(\zeta) = \Psi(x^o + \zeta) + \gamma f(\zeta) \tag{4.129}$$

Necessary conditions for a minimum of Ψ are that the derivative of the Hamiltonian vanish:

$$0 = \frac{\partial H}{\partial \zeta} = \varphi + \Xi \zeta + 2\gamma \Omega \zeta \tag{4.130}$$

and that the constraint be satisfied:

$$\zeta^T \Omega \zeta \le \alpha^2 \tag{4.131}$$

Because the constraint is an inequality, the Lagrange multiplier must satisfy one of the following relations:

$$\gamma \ge 0 \quad \text{if} \quad \zeta^T \Omega \zeta = \alpha^2 \tag{4.132}$$

$$\gamma = 0 \quad \text{if} \quad \zeta^T \Omega \zeta < \alpha^2 \tag{4.133}$$

Eq. (4.130) implies that:

$$\zeta = -\left(\Xi + 2\gamma\Omega\right)^{-1}\varphi \tag{4.134}$$

We must now consider the determination of γ, whose value depends on whether the minimizing value of ζ occurs on the boundary of $Z(\alpha,\omega)$ (eq. (4.132)) or in the interior (eq. (4.133)). From eqs. (4.133) and (4.134) we see that the minimum occurs for a value of ζ in the interior of $Z(\alpha,\omega)$ and thus $\gamma = 0$ if the following inequality holds:

$$\varphi^T\Xi^{-1}\Omega\Xi^{-1}\varphi < \alpha^2 \tag{4.135}$$

If this relation holds then the minimum buckling load is obtained from eq. (4.127) as:

$$\mu(\alpha,\omega) = \Psi(x^o) - \frac{1}{2}\varphi^T\Xi^{-1}\varphi \tag{4.136}$$

On the other hand, if relation (4.135) is not satisfied, then the minimum occurs for a value of ζ on the boundary of $Z(\alpha,\omega)$, and $\gamma \geq 0$. In order to find the value of γ note that eq. (4.132) requires the following equality to hold:

$$\alpha^2 = \zeta^T\Omega\zeta \tag{4.137}$$

which, together with eq. (4.134), determines γ. Then, substituting eq. (4.134) into eq. (4.127) one finds the minimum buckling load to be:

$$\begin{aligned}\mu(\alpha,\omega) &= \Psi(x^o) - \varphi^T\left(\Xi + 2\gamma\Omega\right)^{-1}\varphi \\ &\quad + \frac{1}{2}\varphi^T\left(\Xi + 2\gamma\Omega\right)^{-1}\Xi\left(\Xi + 2\gamma\Omega\right)^{-1}\varphi\end{aligned} \tag{4.138}$$

As in the previous section, a closed-form expression is obtained for the minimum buckling load. Note that if $\Xi \equiv 0$, then this relation reduces to the first-order expression, eq. (4.123). However, when the second-order variation of the buckling with the Fourier coefficients is known, eq. (4.138) enables a more realistic assessment of the minimum buckling than is provided by eq. (4.123).

It should be stressed that the first and second derivatives in eqs. (4.111) and (4.126) may strongly depend on the magnitude of x^o. Moreover, the validity of a truncated Taylor-series approximation appears to us as a challenge both to the numerical analysts and to the experimentalists.

4.3.4 Uniform Bounds on Imperfections

In the previous sections we have represented the initial imperfection profile and its variations exclusively in terms of Fourier coefficients. It turns out to be useful to define the variations of the imperfections in terms of a radial tolerance on the shape of the shell. In this section this tolerance is uniform over the shell, while in the next section the tolerance varies from point to point on the shell.

Let us consider a perfect right circular cylindrical shell of length L and circumference $2\pi R$. A point on the surface is specified by the coordinates y, z where y is the length from one end of the cylinder and z is the distance around the shell from a reference position. Let $\xi = \pi y/L$ and $\theta = z/R$ be normalized positional coordinates. Let $\eta(\xi, \theta)$ be the deviation of a real shell from the perfect cylinder at point (ξ, θ). Let $\eta_o(\xi, \theta)$ be the nominal imperfection profile; it can be chosen as any typical (e.g. average) initial imperfection profile. Let x^o be the significant Fourier coefficients of the nominal spectrum η_o.

In these new circumstances, the allowed variations around η_o are the elements of the following set of functions:

$$H(\hat{\eta}) = \{\eta : \ |\eta(\xi, \theta)| \leq \hat{\eta}\} \tag{4.139}$$

Thus the deviations from the nominal initial imperfection profile are uniformly bounded by the maximum deviation $\hat{\eta}$. In other words, $H(\hat{\eta})$ represents an ensemble of shells for which $\hat{\eta}$ is the (uniform) tolerance in the radial deviation of the shell from its average shape.

For any initial imperfection profile $\eta_o(\xi, \theta) + \eta(\xi, \theta)$, the corresponding vector of significant Fourier coefficients is denoted $x(\eta_o + \eta)$. Let $X(\hat{\eta})$ represent the collection of all Fourier vectors corresponding to shells in the ensemble $H(\hat{\eta})$:

$$X(\hat{\eta}) = \{x : \ x = x(\eta) \quad \text{for} \quad \eta \in H(\hat{\eta})\} \tag{4.140}$$

Also let us define the transposed vector $\varphi^T = \partial \Psi(x^o)/\partial x$ as in eq. (4.112).

With these definitions we can express the non-linear buckling load for first-order deviations from the nominal initial imperfection profile as:

$$\Psi(x^o + x(\eta)) = \Psi(x^o) + \varphi^T \left(x(\eta_o + \eta) - x(\eta_o)\right) \tag{4.141}$$

The Fourier coefficients are linear homogeneous functions of the imperfection profile. This means that:

$$x(\eta_o + \eta) - x(\eta_o) = x(\eta) \tag{4.142}$$

Consequently eq. (4.141) becomes:

$$\Psi(x^o + x(\eta)) = \Psi(x^o) + \varphi^T x(\eta) \tag{4.143}$$

Now we can determine the lowest buckling load obtainable for any uniformly bounded deviation $\eta(\xi, \theta)$ from the initial nominal imperfection profile η_o :

$$\mu(\hat{\eta}) = \min_{x(\eta) \in X(\hat{\eta})} \left(\Psi(x^o) + \varphi^T x(\eta) \right) \tag{4.144}$$

$$= \Psi(x^o) + \min_{\eta \in H(\hat{\eta})} \varphi^T x(\eta) \tag{4.145}$$

$\mu(\hat{\eta})$ is the lowest buckling load of any shell in the ensemble of shells defined by the set $H(\hat{\eta})$. Eq. (4.145) calls for the minimum of the linear functional $\varphi^T x(\eta)$ on the convex set $H(\hat{\eta})$. This extremum can be sought on the set $T(\hat{\eta})$ of extreme-point functions of $H(\hat{\eta})$. Thus:

$$\mu(\hat{\eta}) = \Psi(x^o) + \min_{\eta \in T(\hat{\eta})} \varphi^T x(\eta) \tag{4.146}$$

Let D represent the domain of the surface of the cylinder:

$$D = \{(\xi, \theta) : \ 0 \leq \xi \leq \pi \quad , \quad 0 \leq \theta \leq 2\pi\} \tag{4.147}$$

The set of extreme-point functions is:

$$T(\hat{\eta}) = \{\eta : \ \eta(\xi, \theta) = \hat{\eta}\left(K_P(\xi, \theta) - K_Q(\xi, \theta)\right), \quad P \cap Q = \emptyset \ , \ P \cup Q = D\} \tag{4.148}$$

where $K_V(\xi, \theta)$ is a characteristic function, defined as follows: For any set V of points on the surface of the shell, $K_V(\xi, \theta) = 1$ if the point (ξ, θ) belongs to the set V, and equals zero otherwise. Thus $T(\hat{\eta})$ is the set of all functions $\eta(\xi, \theta)$ which switch arbitrarily back and forth between $+\hat{\eta}$ and $-\hat{\eta}$ as (ξ, θ) varies over the surface of the shell.

We will now proceed to evaluate the minimum in eq. (4.146). Before doing so we need to evaluate the Fourier coefficients of an arbitrary element of $T(\hat{\eta})$. We can approximate the initial imperfection function in a truncated two-dimensional Fourier series as follows:

$$\eta(\xi,\theta) = \sum_{i=0}^{N_1} a_i \cos i\xi + \sum_{j=1}^{N_2}\sum_{k=1}^{N_3}(b_{jk}\sin j\xi\cos k\theta + c_{jk}\sin j\xi\sin k\theta) \quad (4.149)$$

The coefficients in this expansion are evaluated as:

$$a_i(\eta) = \frac{1}{(1+\delta_{i0})\pi^2}\int_0^{2\pi}\int_0^{\pi}\eta(\xi,\theta)\cos i\xi\, d\xi\, d\theta \quad i = 0,1,2,\ldots \quad (4.150)$$

$$b_{jk}(\eta) = \frac{2}{\pi^2}\int_0^{2\pi}\int_0^{\pi}\eta(\xi,\theta)\sin j\xi\cos k\theta\, d\xi\, d\theta \quad j,k>0 \quad (4.151)$$

$$c_{jk}(\eta) = \frac{2}{\pi^2}\int_0^{2\pi}\int_0^{\pi}\eta(\xi,\theta)\sin j\xi\sin k\theta\, d\xi\, d\theta \quad j,k>0 \quad (4.152)$$

where δ_{i0} is the Kronecker delta function.

To develop an explicit expression for the minimum buckling load, let us adopt the following nomenclature for the elements of the vector φ:

$$\alpha_i = \frac{\partial\Psi(x^o)}{\partial a_i}, \quad i = 0,1,2,\ldots \quad (4.153)$$

$$\beta_{jk} = \frac{\partial\Psi(x^o)}{\partial b_{jk}}, \quad j,k>0 \quad (4.154)$$

$$\gamma_{jk} = \frac{\partial\Psi(x^o)}{\partial c_{jk}}, \quad j,k>0 \quad (4.155)$$

Expanding the inner product $\varphi^T x(\eta)$ which appears in eq. (4.143), explicitly in terms of the quantities α_i, β_{jk}, γ_{jk}, $a_i(\eta)$, $b_{jk}(\eta)$ and $c_{jk}(\eta)$, one obtains the following expression for the buckling load, to first order in the imperfection profile:

$$\Psi(x^o + x(\eta)) = \Psi(x^o) + \sum_{i=0}^{N_1}\alpha_i a_i(\eta) + \sum_{j=1}^{N_2}\sum_{k=1}^{N_3}(\beta_{jk}b_{jk}(\eta) + \gamma_{jk}c_{jk}(\eta)) \quad (4.156)$$

Let us define the following function:

$$S(\xi,\theta) = \frac{1}{\pi^2}\sum_{i=0}^{N_1}\frac{1}{1+\delta_{i0}}\alpha_i \cos i\xi + \frac{2}{\pi^2}\sum_{j=1}^{N_2}\sin j\xi \sum_{k=1}^{N_3}(\beta_{jk}\cos k\theta + \gamma_{jk}\sin k\theta) \tag{4.157}$$

Combining eqs. (4.150) – (4.152) with eq. (4.156) one finds that the buckling load for an initial imperfection spectrum η is given by:

$$\Psi(x^o + x(\eta)) = \Psi(x^o) + \int_0^{2\pi}\int_0^{\pi} \eta(\xi,\theta)S(\xi,\theta)d\xi d\theta \tag{4.158}$$

We now wish to determine the lowest value of this buckling load for the ensemble $H(\hat{\eta})$ of shells. Thus we seek the minimum of $\Psi(x^o + x(\eta))$ as η varies on $H(\hat{\eta})$. As we mentioned in connection with eq. (4.145), this minimization may be sought as η varies on the set $T(\hat{\eta})$ of extreme-point functions. Examination of eq. (4.148) shows that each extreme-point function $\eta(\xi,\theta)$ switches between $+\hat{\eta}$ and $-\hat{\eta}$ as (ξ,θ) moves over the domain D. Thus eq. (4.158) is minimized by choosing $\eta = +\hat{\eta}$ for those values of (ξ,θ) at which $S(\xi,\theta)$ is negative, and by choosing $\eta = -\hat{\eta}$ at those values of (ξ,θ) for which $S(\xi,\theta)$ is positive. Equivalently, the minimum buckling load is obtained from that function $\eta(\xi,\theta) \in T(\hat{\eta})$ whose sign is always the opposite of the sign of $S(\xi,\theta)$. Thus for the minimum buckling load we arrive at:

$$\mu(\hat{\eta}) = \Psi(x^o) - \hat{\eta}\int_0^{2\pi}\int_0^{\pi} |S(\xi,\theta)|d\xi d\theta \tag{4.159}$$

As anticipated, this formula indicates that the minimum buckling load of the ensemble of shells with uncertain but uniformly bounded imperfections is lower than for the nominal shell. The decrease of the buckling load can be readily estimated by eq. (4.159). Using data from Elishakoff *et al* (1987) one finds the integral equal to 0.4555. Thus the minimum buckling load for an ensemble whose imperfections are uniformly bounded by $\hat{\eta}$ is $\mu = 0.746 - 0.4555\hat{\eta}$. For example, $\mu = 0.700$ for $\hat{\eta} = 0.1$. That is, if the uniform bound on the initial imperfections constitutes one tenth of the shell thickness, then the buckling load of the weakest shell in the ensemble is 70 percent of the classical buckling load. These numerical results should be viewed with caution, as they are based on an incomplete set of partial derivatives of the Ψ function. Derivatives of Ψ with respect to additional imperfection modes may significantly alter the numerical value of the minimum buckling load. The calculations presented here demonstrate the feasibility of this analysis.

4.3.5 Envelope-Bounds on Imperfections

The derivation of eq. (4.158) depends on the convexity of the set of initial imperfection functions, and on the Taylor expansion in eq. (4.143) being to first order, but not on the specific structure of the initial imperfection set. We are thus free to employ eq. (4.158) for a first-order analysis when the function $\eta(\xi,\theta)$ is an extreme-point function of *any* convex set of initial imperfection functions. A useful generalization of the uniform-bound model of initial imperfections is to consider imperfection profiles which are contained in an envelope. Let us consider the following set of initial imperfection functions:

$$H(\eta_l,\eta_u) = \{\eta : \eta_l(\xi,\theta) \le \eta(\xi,\theta) \le \eta_u(\xi,\theta)\} \tag{4.160}$$

The set $H(\eta_l,\eta_u)$ represents an ensemble of shells for which the radial tolerance varies over the surface of the shell; $\eta_l(\xi,\theta)$ is the lower envelope function and $\eta_u(\xi,\theta)$ is the upper envelope function. The extreme-point functions of $H(\eta_l,\eta_u)$ are the functions belonging to the set:

$$\begin{aligned} T(\eta_l,\eta_u) &= \{\eta : \eta(\xi,\theta) = \eta_u(\xi,\theta)K_P(\xi,\theta) + \eta_l(\xi,\theta)K_Q(\xi,\theta), \\ & \qquad P\cap Q = \emptyset, P\cup Q = D\} \end{aligned} \tag{4.161}$$

where D represents the domain of the shell surface and is defined in eq. (4.147). The value of a function in $T(\eta_l,\eta_u)$ switches between $\eta_l(\xi,\theta)$ and $\eta_u(\xi,\theta)$ as (ξ,θ) moves over the domain D.

Examination of eq. (4.158) reveals that the minimum buckling load is obtained for that function which takes the lower value, $\eta_l(\xi,\theta)$, when $S(\xi,\theta)$ is positive and takes the upper value, $\eta_u(\xi,\theta)$, when $S(\xi,\theta)$ is negative.

To conveniently formulate the minimum buckling load, let us define the following two subsets of D :

$$\Delta_+ = \{(\xi,\theta) : \; S(\xi,\theta) \ge 0\} \tag{4.162}$$

$$\Delta_{-} = \{(\xi,\theta) : \; S(\xi,\theta) < 0\} \tag{4.163}$$

Thus Δ_{+} is the set of points (ξ,θ) on the surface of the shell for which the function $S(\xi,\theta)$ is non-negative. Likewise, Δ_{-} is the set of surface points at which $S(\xi,\theta)$ is negative.

The lowest buckling load, to first order in the imperfection profile, obtainable from any initial imperfection profile bounded within the envelope defined in eq. (4.160) is:

$$\mu(\eta_l,\eta_u) = \Psi(x^o) + \int_{\Delta_{-}} \eta_u(\xi,\theta)S(\xi,\theta)\,d\xi\,d\theta + \int_{\Delta_{+}} \eta_l(\xi,\theta)S(\xi,\theta)\,d\xi\,d\theta \tag{4.164}$$

This relation has several practical implications. First of all, one realizes that the upper bound, η_u, on the ensemble of initial imperfection profiles influences the value of the minimum buckling load only in the domain Δ_{-}. That is, η_u can assume any values whatsoever[5] in Δ_{+} without altering the buckling load of the weakest shell in the ensemble $H(\eta_l,\eta_u)$. Similarly, the lower bound, η_l, effects the buckling load only at points in Δ_{+}, and can be freely chosen[6] in Δ_{-}.

Following this line of thought, it is convenient to characterize the ensemble $H(\eta_l,\eta_u)$ of shells with a single tolerance function, $\tau(\xi,\theta)$, rather than with two envelope functions η_l and η_u. Let $\tau(\xi,\theta)$ be the function:

$$\tau(\xi,\theta) = \begin{cases} \eta_u(\xi,\theta) & \text{for} \quad (\xi,\theta) \in \Delta_{-} \\ \eta_l(\xi,\theta) & \text{for} \quad (\xi,\theta) \in \Delta_{+} \end{cases} \tag{4.165}$$

Let $H'(\tau)$ be the set of initial imperfection profiles, η, which satisfy:

$$\eta(\xi,\theta) \geq \tau(\xi,\theta) \;\; \text{for} \;\; (\xi,\theta) \in \Delta_{+} \tag{4.166}$$

$$\eta(\xi,\theta) \leq \tau(\xi,\theta) \;\; \text{for} \;\; (\xi,\theta) \in \Delta_{-} \tag{4.167}$$

The sets $H(\eta_l,\eta_u)$ and $H'(\tau)$ are not identical. However, the buckling loads of the weakest shell in each of these ensembles are the same. In other words, τ (and eqs. (4.166) and 4.167)) is equivalent to η_l and η_u (and eq. (4.160)) as far as the minimum buckling load is concerned.

[5]Subject to two important restrictions: That $\eta_u \geq \eta_l$ throughout D, and that the magnitude of η_u not become so large as to invalidate the use of a first-order expansion of $\Psi(x^o + x(\eta))$.

[6]Subject to the same two constraints, applied to η_l.

The tolerance function is defined in terms of the sign of $S(\xi, \theta)$. The magnitude of S also carries a physical significance, and can be thought of as a measure of the sensitivity to imperfection in the infinitestimal portion $d\xi d\theta$ of the shell at point ξ, θ. The magnitude of S at each point on the surface of the shell serves to weight the local contribution to the buckling load of an initial imperfection at that point. As indicated by eqs. (4.158) and (4.164), if the magnitude of S is small over a region of the surface, then the initial imperfections in that region can be comparatively large without excessively enlarging the buckling load. Conversely, the buckling load is very sensitive to imperfections in those regions of the shell for which $|S|$ is large. The axial variation of S is demonstrated in fig. 4.16, based on data from Elishakoff *et al* (1987). This figure illustrates that the contribution to imperfection-sensitivity of the top and bottom portions of the shell is near zero, while around the midplane the sensitivity achieves its maximum values. Figures such as this provide useful insight to the spatially varying sensitivity of the shell to initial imperfections.

A tolerance function defines an ensemble of shells, as in eqs. (4.166) and (4.167). Suppose that one wishes to construct a radial tolerance function for which the minimum buckling load on the corresponding ensemble assumes the value M. That is, one wishes to choose τ so that:

$$M = \Psi(x^o) + \int_0^{2\pi}\int_0^{\pi} \tau(\xi, \theta) S(\xi, \theta)\, d\xi\, d\theta \tag{4.168}$$

Furthermore, suppose one desires the tolerance in each region to be as large as possible, consistent with eq. (4.168). One way to do this is to choose the tolerance so that the local contribution to the minimum buckling load is uniform over the surface of the shell. This requires that:

$$\tau(\xi, \theta) S(\xi, \theta) = \text{constant} \tag{4.169}$$

Combining these two equations yields the following expression for the desired radial tolerance function:

$$\tau(\xi, \theta) = \frac{M - \Psi(x^o)}{2\pi^2 S(\xi, \theta)} \tag{4.170}$$

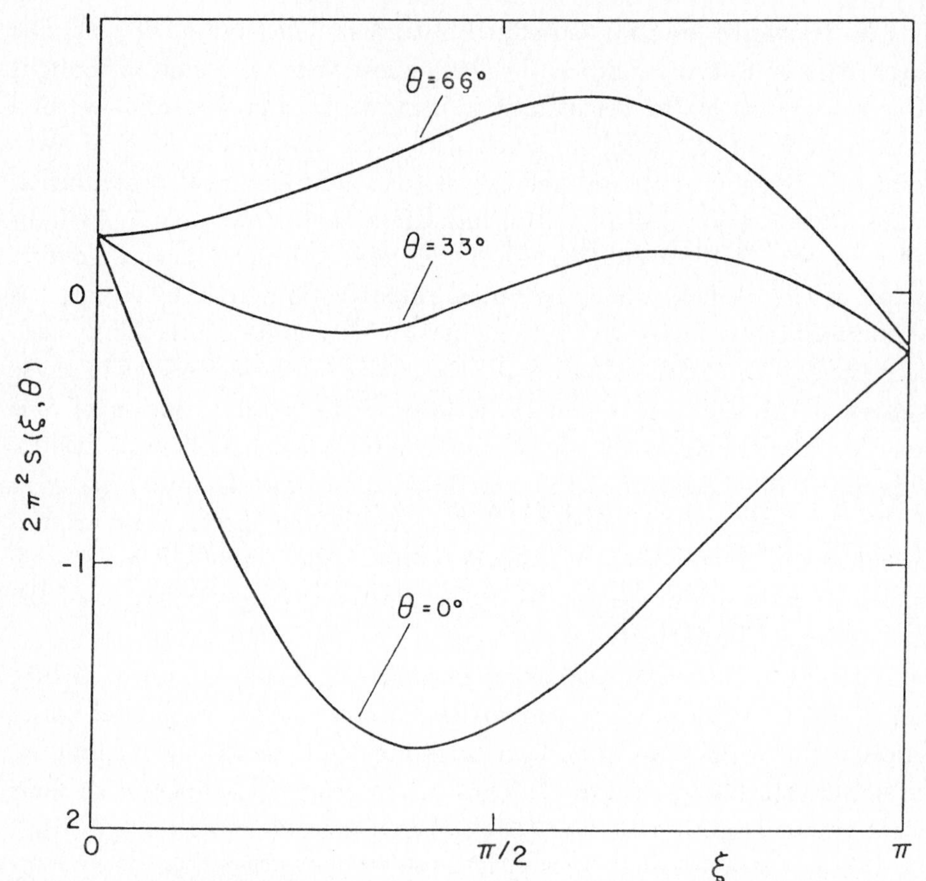

Figure 4.16: Axial variation of the local sensitivity to imperfection, for three different azimuthal angles.

Note that the sign of τ is always the opposite of the sign of S, because the nominal buckling load, $\Psi(x^o)$, exceeds the ensemble minimum M. This is consistent with the use of eq. (4.170) as a tolerance function, as defined in eqs. (4.166) and (4.167).

To summarize, envelope-bound models allow one to study the effect (on the minimum buckling load) of relaxing the radial tolerance selectively in different areas of the shell. The function $S(\xi, \theta)$ enables one to assess the contribution to imperfection-sensitivity of various portions of the shell, and makes it possible to design a spatially varying tolerance function accordingly.

4.3.6 Estimates of the Knockdown Factor

The knockdown factor κ is an engineering parameter whose product with the classical buckling load, P_{cl}, yields a lower bound for the buckling load of the structure. The knockdown factor can be viewed as a property of an ensemble of shells, and the lower bound is the buckling load of the weakest shell in the ensemble. In the previous sections we have obtained explicit expressions for the minimum buckling load μ of an ensemble of shells. The ratio μ/P_{cl} provides an estimate of the knockdown factor. This estimate can be evaluated for each of the models which has been discussed, thereby relating the knockdown factor to different characterizations of the uncertainty in the initial imperfections.

It will be noted that μ depends on the choice of a nominal initial imperfection spectrum, x^o. If the initial imperfection spectra of the ensemble tend to cluster around an average spectrum, $\bar{x}$, then a reasonable estimate of the knockdown factor would be:

$$\kappa = \frac{\mu(\bar{x})}{P_{cl}} \tag{4.171}$$

A noteworthy characteristic of this relation is that it enables estimation of the knockdown factor based on limited empirical knowledge of ensemble: The mean imperfection of the ensemble and the tolerance to which the ensemble was produced.

In this section we have considered the buckling of shells with general, geometrical imperfections. Instead of assuming extensive knowledge of the probabilistic characteristics of the initial imperfections, we have assumed that the initial imperfections are uncertain but bounded. Three different convex models of uncertainty have been studied. The set $Z(\alpha, \omega)$ (eq. (4.113)) represents an ensemble of shells whose Fourier coefficients are contained in an ellipsoid. The set $H(\hat{\eta})$ (eq. (4.139)) defines an ensemble for which $\hat{\eta}$ is a uniform bound on the tolerance in the radial dimension of the shell. Finally, $H(\eta_l, \eta_u)$ (eq. (4.160)) represents an ensemble of shells whose manufacture is subject to a radial tolerance which varies over the surface of the shell. For each of these ensembles we have obtained an explicit expression for the buckling load of the weakest shell in the ensemble, and we have related this to the

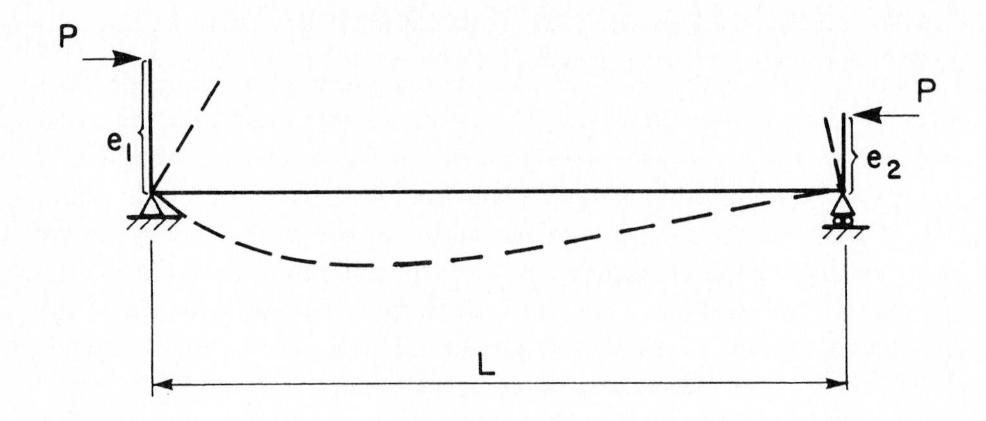

Figure 4.17: Elastic bar subject to axial compression with uncertain eccentricity

knockdown factor. We have shown that the function $S(\xi, \theta)$ can be used to define a spatially varying radial tolerance. Finally, we have demonstrated that numerical results from sophisticated non-linear buckling codes can be readily incorporated in the evaluation of these quantities.

4.3.7 First- and Second-Order Analyses

In the foregoing section we developed first-order and second-order analyses for the imperfection-sensitivity of cylindrical shells. An explicit expression of the buckling load, in terms of the imperfection parameters, was unavailable. It is convenient, therefore, to compare first- and second-order approximations in a situation for which an analytical expression for the response to uncertain parameters is available. We will consider an elastic bar compressed by an axial force. The uncertainty is described by the eccentricities e_1 and e_2 of the application of the axial compression (Fig. 4.17).

The differential equation describing the deflection of the bar reads:

$$EI\frac{d^4w}{dx^4} + P\frac{d^2w}{dx^2} = 0 \ , \quad 0 \le x \le L \tag{4.172}$$

The authors are indebted to Yossi Ganashvili for assistance in performing the calculations presented in this section.

where EI is the flexural stiffness, P is the axial compression, w is the displacement and L is the length of the bar. Let us denote:

$$k^2 = \frac{P}{EI} \tag{4.173}$$

The general solution of eq. (4.172) is:

$$w(x) = C_1 \sin kx + C_2 \cos kx + C_3 x + C_4 \tag{4.174}$$

The bending moment of the bar is:

$$M_z(x) = -EIw''(x) = P(C_1 \sin kx + C_2 \cos kx) \tag{4.175}$$

The boundary conditions in terms of the bending moments are:

$$M_z(0) = Pe_1 \quad , \quad M_z(L) = Pe_2 \tag{4.176}$$

which yields the following final expression for the bending moment:

$$M_z(x) = \frac{P}{\sin kL}(e_2 - e_1 \cos kL) \sin kx + Pe_1 \cos kx \tag{4.177}$$

We are interested in the maximum bending moment over the length of the bar. Consider the function:

$$y(x) = A \sin kx + B \cos kx \tag{4.178}$$

The maximum value of $y(x)$ is found to be:

$$y_{\max} = \sqrt{A^2 + B^2} \tag{4.179}$$

Hence, the spacewise maximum bending moment M_z^* in the bar is:

$$M_z^*(e_1, e_2) = \frac{P}{\sin kL}\sqrt{e_1^2 + e_2^2 - 2e_1 e_2 \cos kL} \tag{4.180}$$

This expression coincides with Eq. 1.44 in the book of Pikovsky (1961, p52) (see also Young (1932, p507) and Timoshenko (1963, p42)).

The position x^* at which the bending moment reaches its maximum satisfies:

$$(e_2 - e_1 \cos kL) \cos kx^* - e_1 \sin kL \sin kx^* = 0 \tag{4.181}$$

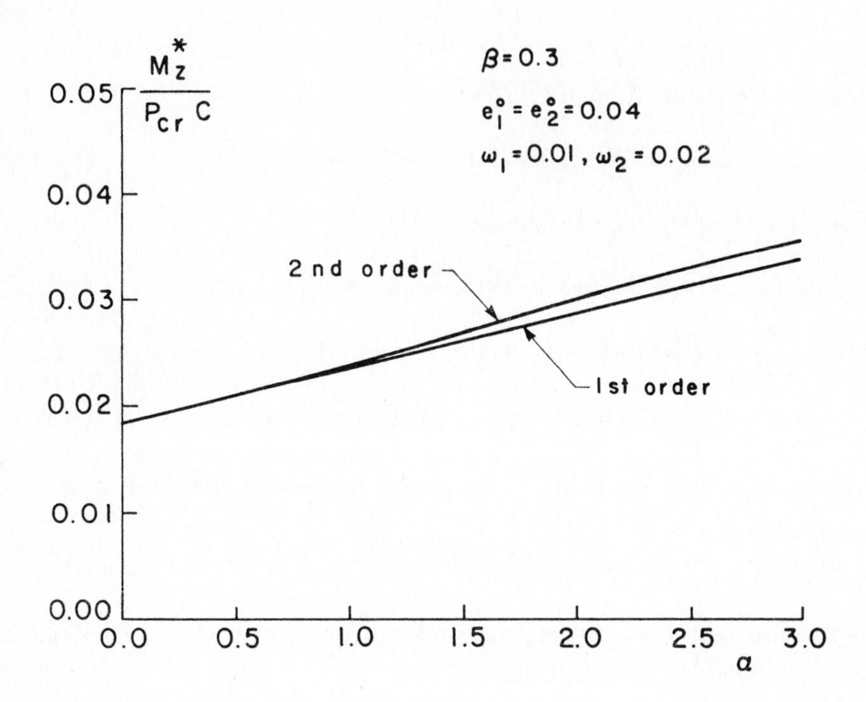

Figure 4.18: Maximum bending moment $\mu(\alpha, \omega_1, \omega_2)$ versus α, for dimensionless load ratio $\beta = 0.3$

or

$$\tan kx^* = \left(\frac{e_2}{e_1} - \cos kL\right)\frac{1}{\sin kL} \tag{4.182}$$

In order for this maximum to take place in the interior of the column, $0 < x^* < L$, the following condition must be fulfilled (Pikovsky, 1961):

$$\cos kL < \frac{e_2}{e_1} < \frac{1}{\cos kL} \tag{4.183}$$

for

$$0 < P < \frac{\pi^2 EI}{4L^2} \quad , \quad 0 < kL < \frac{\pi}{2} \tag{4.184}$$

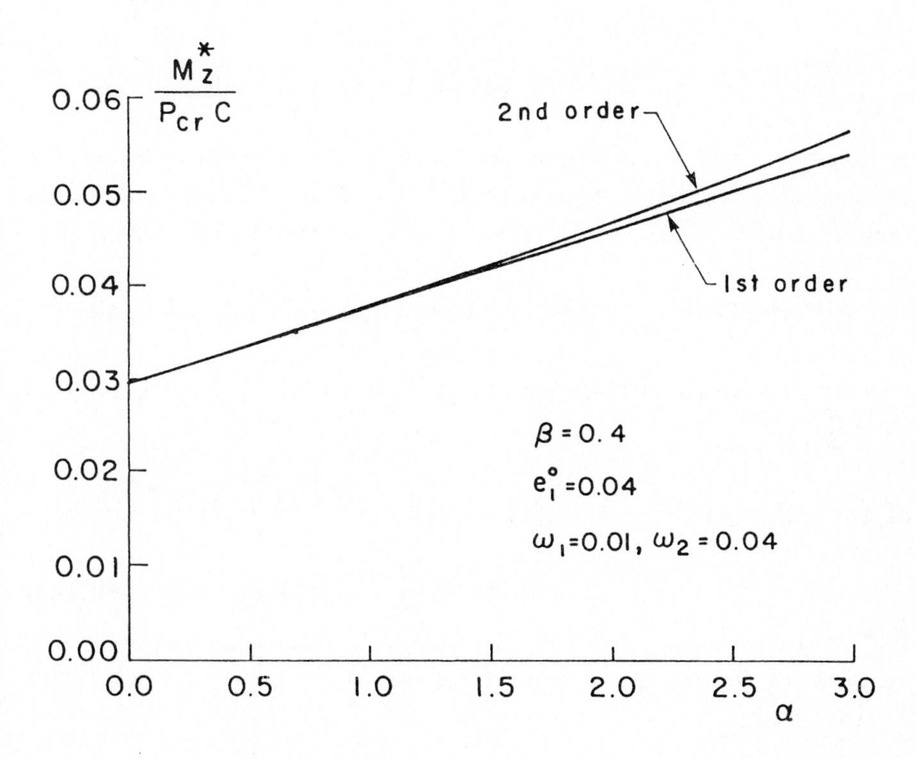

Figure 4.19: Maximum bending moment $\mu(\alpha, \omega_1, \omega_2)$ versus α, for dimensionless load ratio $\beta = 0.4$

Otherwise the maximum bending moment is reached at the ends of the bar and are either Pe_1 or Pe_2.

For

$$\frac{\pi^2 EI}{4L^2} < P < \frac{\pi^2 EI}{L^2} \tag{4.185}$$

the conditions in (4.183) are not needed.

Once e_1 and e_2 are specified, the maximum value of the bending moment can be directly evaluated from the above analysis. Assume now that the initial eccentricities are uncertain. The nominal values of the eccentricities are e_1^o and e_2^o respectively, and the deviations from these nominal values are ζ_1 and ζ_2, respectively. We will assume that these deviations vary within the ellipsoidal set:

$$Z(\alpha, \omega_1, \omega_2) = \left\{ (\zeta_1, \zeta_2) : \ \left(\frac{\zeta_1}{\omega_1}\right)^2 + \left(\frac{\zeta_2}{\omega_2}\right)^2 \leq \alpha^2 \right\} \tag{4.186}$$

where again ω_1 and ω_2 are semi-axes, and α is a size parameter. We are interested in finding the maximum, with respect to the uncertainty in the eccentricity, of the spacewise maximum bending moment:

$$\mu(\alpha, \omega_1, \omega_2) = \max_{\zeta_1, \zeta_2 \in Z(\alpha, \omega_1, \omega_2)} M_z^*(e_1 + \zeta_1, e_2 + \zeta_2) \tag{4.187}$$

In accordance with the first-order analysis, we replace $M_z^*(e_1+\zeta_1, e_2+\zeta_2)$ by:

$$\begin{aligned} M_z^*(e_1 + \zeta_1, e_2 + \zeta_2) &= M_z^*(e_1^o, e_2^o) + \frac{\partial M_z^*(e_1 + \zeta_1, e_2 + \zeta_2)}{\partial e_1} \zeta_1 \\ &+ \frac{\partial M_z^*(e_1 + \zeta_1, e_2 + \zeta_2)}{\partial e_2} \zeta_2 \end{aligned} \tag{4.188}$$

Then, without repeating the analysis given in Section 4.3.2, we arrive at the expression:

$$\begin{aligned} \mu(\alpha, \omega_1, \omega_2) &= M_z^*(e_1^o, e_2^o) \\ &+ \alpha \sqrt{\left(\omega_1 \frac{\partial M_z^*(e_1^o, e_2^o)}{\partial e_1}\right)^2 + \left(\omega_2 \frac{\partial M_z^*(e_1^o, e_2^o)}{\partial e_2}\right)^2} \end{aligned} \tag{4.189}$$

For convenience of notation, let us define:

$$\beta_1 = e_1 - e_2 \cos kL \tag{4.190}$$

$$\beta_2 = e_2 - e_1 \cos kL \tag{4.191}$$

$$\beta_3 = e_1^2 + e_2^2 - 2e_1 e_2 \cos kL \tag{4.192}$$

We obtain the following explicit expressions for the derivatives in eq. (4.189).

$$\frac{\partial M_z^*}{\partial e_1} = \frac{P\beta_1}{\sqrt{\beta_3} \sin kL} \tag{4.193}$$

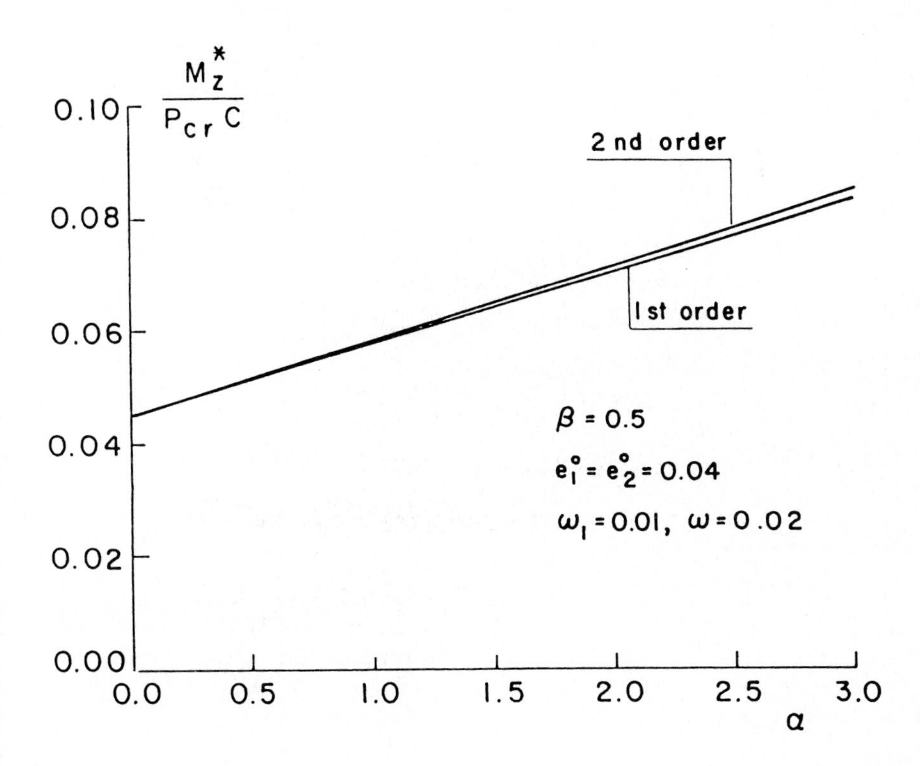

Figure 4.20: Maximum bending moment $\mu(\alpha,\omega_1,\omega_2)$ versus α, for dimensionless load ratio $\beta = 0.5$

$$\frac{\partial M_z^*}{\partial e_2} = \frac{P\beta_2}{\sqrt{\beta_3}\sin kL} \tag{4.194}$$

The second-order analysis yields a formula analogous to eq. (4.138):

$$\begin{aligned}\mu(\alpha,\omega_1,\omega_2) &= M_z^*(e_1^o, e_2^o) + \phi^T(\Xi + 2\gamma\Omega)^{-1}\phi \\ &+ \frac{1}{2}\phi^T(\Xi + 2\gamma\Omega)^{-1}\Xi(\Xi + 2\gamma\Omega)^{-1}\phi\end{aligned} \tag{4.195}$$

The matrix Ξ is symmetric with the following elements:

$$\xi_{11} = \frac{\partial^2 M_z^*}{\partial e_1^2} = \frac{P\,(\beta_3 - \beta_1^2)}{\beta_3^{3/2}\sin kL} \tag{4.196}$$

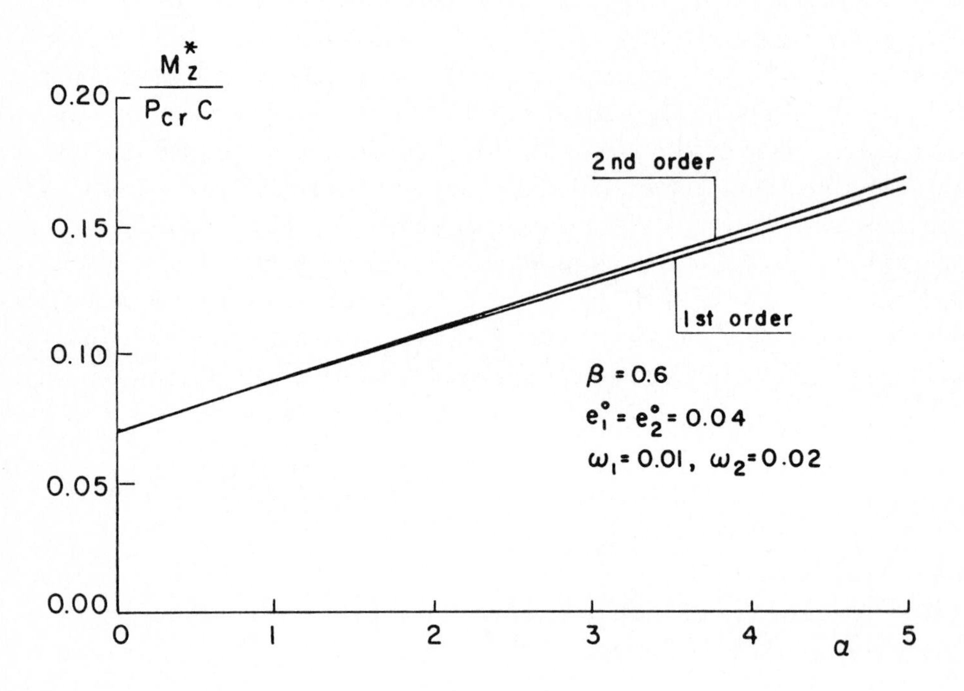

Figure 4.21: Maximum bending moment $\mu(\alpha, \omega_1, \omega_2)$ versus α, for dimensionless load ratio $\beta = 0.6$

$$\xi_{12} = \xi_{21} = \frac{\partial^2 M_z^*}{\partial e_1 \partial e_2} = -\frac{P\left(\beta_3 \cos kL + \beta_1(e_2 - 2e_1 \cos kL)\right)}{\beta_3^{3/2} \sin kL} \tag{4.197}$$

$$\xi_{22} = \frac{\partial^2 M_z^*}{\partial e_2^2} = \frac{P\left(\beta_3 - \beta_2^2\right)}{\beta_3^{3/2} \sin kL} \tag{4.198}$$

Results of calculations in accordance with the first-order and the second-order analyses are portrayed in Figs. 4.18 – 4.21. All figures are associated with values $e_1^o = e_2^o = 0.04$, $\omega_1 = 0.01$ and $\omega_2 = 0.02$. In

Fig. 4.18 the dimensionless load $\beta = P/P_{cl}$, where $P_{cl} = \pi^2 EI/L^2$ is the Euler buckling load, is fixed at $\beta = 0.3$, in Figs. 4.19 - 4.21 this value is varied between 0.4 to 0.6. c denotes the radius of inertia of the cross-section of the bar. The maximum moment increases with the size, α, of the uncertainty ellipsoid. For moderate values of the size parameter α the agreement between the first-order and second-order analyses is excellent. In comparing Figs. 4.18 - 4.21 one notes that, with the increase of the nondimensional applied load, the percentage-wise disagreement between the first-order and the second-order analyses decreases considerably. We conclude therefore that the first-order analysis yields acceptable results for small uncertainties and greater loads. One should be warned, however, that the similarity of the first- and second-order approximations that occurs for the elastic bar may not be characteristic of more complicated structures like plates and shells.

Chapter 5

Concluding Remarks

> We mathematicians who operate with nothing more expensive than paper and possibly printer's ink are quite reconciled to the fact that, if we are working in a very active field, our discoveries will commence to be obsolete at the moment they are written down or even at the moment they are conceived. We know that for a long time everything we do will be nothing more than the jumping-off point for those who have the advantage of already being aware of our ultimate results. This is the meaning of the famous apothegm of Newton, when he said, "If I have seen further than other men, it is because I have stood on the shoulders of giants."[1]

A mathematical model of a natural or technological system embodies quantitative knowledge of the properties of the system. However a model, like a painting, expresses the taste, goals, understanding and perspective of its creator. The representation of a given system is rarely limited to only a single possible model, and greater system-complexity implies wider diversity of models. Uncertain systems are usually of great complexity by virtue of their myriad possible realizations. Models of such systems consequently show wide variability. The engineer generally selects his models for their utility.

The theory of probability has proven extraordinarily useful in modelling uncertainty in every branch of engineering. Probability is here

[1] Wiener, 1956, pp 266 – 267.

to stay, and justly so. However, the information content of a probabilistic model is often quite high, which means that implementation of a realistic model requires extensive data and thorough scientific understanding of the system being modelled. In chapter 1 we tried to demonstrate that the resolution of analysis and design problems can be sensitive to those parts of a probabilistic model which are most difficult to establish precisely.

Subjective judgements are also an integral part of design, and probabilistic models generate the need for a certain type of qualitative decision-making. The concept of event-likelihood pervades decisions which are made on the basis of a probabilistic model. For example, the probabilistic concepts of reliability, of false-alarm rate or of failed detection, when used as design tools, all require the engineer to establish an acceptable probability of failure, or of false alarm or of failed detection. These are sometimes difficult decisions. A different model may avoid these questions, although other subjective decisions must of course be made instead.

The convex models described in this book are quite different from probabilistic ones. As a rule, convex models require only fragmentary knowledge of the system being modelled. Also, the subjective design decisions which are generated by convex models are divorced from the concept of chance. We hope to have demonstrated that, despite their simplicity, convex models are useful in a range of analysis and design problems which arise in applied mechanics.

The information content of a convex model is usually a bound or constraint which defines an infinite set of values or functional forms which an uncertain variable may assume. Commonly encountered convex models are based on the following types of constraints.

1. Envelope bounds:

$$f_{\min}(x) \leq f(x) - \bar{f}(x) \leq f_{\max}(x) \tag{5.1}$$

where $f_{\min}(x)$ and $f_{\max}(x)$ are known functions, chosen to delimit the range of deviation of $f(x)$ from its nominal form, $\bar{f}(x)$. For example, localization of deviations to a particular domain X of values of x is represented by an envelope-bound convex model in which $f_{\min}(x)$ and $f_{\max}(x)$ vanish outside of X. Envelope-bound models, as well as the important special case of uniform-bounds, have found application throughout this book.

2. Energy bounds:

$$\int \left(f(x) - \bar{f}(x)\right)^2 dx \leq E \tag{5.2}$$

where E is the "energy" of deviation of $f(x)$ from its nominal value, $\bar{f}(x)$. Energy-bound convex models are useful for representing transient phenomena of possibly unbounded magnitude.

3. Ellipsoidal Fourier bounds:

$$\varphi^T W \varphi \leq 1 \tag{5.3}$$

where φ is a vector of Fourier coefficients of the uncertain function and W is a weighting matrix. In chapter 4 we found ellipsoidal Fourier bounds useful for describing geometric imperfections of structures.

Other convex models are also available, as described in chapter 2, such as sets of integral-bounded or monotonic or unimodal functions. Furthermore, constraints can be applied to derivatives of the uncertain functions, thereby generating slope-bounded or curvature-bounded convex models of all sorts. For example, slope-bounded convex models were employed in sections 3.2.6 and 3.2.7 in the analysis of vehicle motion on an uncertain obstacle. Convex models can be formulated for vector valued functions as well as for scalars.

The inequality constraints which define a convex model generate subjective design decisions characterized by reference to extremal situations. For example, we found in section 3.2 that the suspension system of a vehicle moving on a rough surface can be designed to assure that the vertical acceleration of the center of mass does not exceed a given maximum. The designer must independently establish the greatest acceptable value of acceleration. In section 3.4 we investigated the design of vibration measurements for characterizing unknown uniformly bounded excitations of an elastic system. The designer must decide what degree of resolution is required in characterizing the excitation. The analysis then indicates if the desired resolution is attainable and, if so, what measurements are required. In section 4.3 we found that the radial tolerance for manufacture of thin cylindrical shells subject to geometrical imperfections can be chosen to assure that the buckling load of the weakest shell does not fall below a given minimum. The designer must decide what minimum buckling load can be tolerated.

The convex models and their applications to engineering systems discussed in this book do not exhaust the field of possibilities. It is fair to say, however, that most extensions are presently in the realm of active research. Following is a brief exposition of subjects which seem ripe for investigation.

Multiplicative Uncertainty. Consider a dynamical system whose state vector, $x(t)$, is described by the following differential equation:

$$\frac{dx}{dt} = A(t)x(t) \tag{5.4}$$

This system is subject to multiplicative uncertainty if the dynamics matrix, $A(t)$, is uncertain. It may be possible to formulate a convex model for the uncertainty of $A(t)$ by specifying a convex set $\mathcal{A}$ of allowed dynamics matrices. The problems of analysis and design associated with multiplicative uncertainty are different from those considered in this book because, though $\mathcal{A}$ is convex, the set of responses will almost invariably be non-convex because of the non-linear, multiplicative, relation between x and A. Multiplicative uncertainty can arise from modelling errors or from system faults which cause modification of the dynamics matrix. These modifications can arise in mechanical systems through alteration of the mass, stiffness or damping matrices.

Non-Linear Systems. Multiplicative uncertainty is a special case of uncertain non-linear systems. Nature can supply a seemingly inexhaustible variety of non-linearities. The challenge in each case will be to perform the analysis or design by exploiting both the structure of the non-linearity and the convexity of the set of uncertain variables.

Algorithmic Problems. The analysis and design of complex systems leads to numerical computation, usually constrained optimization. For the problems discussed in this book these are often linear optimizations on convex sets. More complicated problems requiring special algorithmic treatment will arise in connection with non-linear systems.

Applications. Numerous applications remain to be studied, each of which may be expected to generate new questions. We mention just a few fields which are subject to complex sources of uncertainty and amenable to analysis with convex models.

1. Diagnosis and control of failure in dynamical systems, as discussed in connection with multiplicative uncertainty.

2. Signal processing and filtering with complex noise sources for which probabilistic models are unavailable. For example, complex uncertain disturbances are often encountered in acoustic or vibrational monitoring of machine tool operation (Braun, 1986).

3. Two-phase flow, where probabilistic description of physical parameters such as heat conduction coefficients has proven quite difficult.

4. Stress concentration due to uncertain irregularities in the shape of holes in structures, which may enhance the stress concentration factors.

The reader is invited to extend this list as well as to tackle those problems of his choice. We believe that the concepts and techniques put forward in this monograph will be useful, both in problem solving and in broadening the engineer's thinking about the modelling of uncertainty.

Bibliography

[1] Adams, H., *The Education of Henry Adams,* E. Samuels, ed., Houghton Mifflin Company, Boston, (1918, 1973).

[2] Akerhielm, F., Espefalt, R. and Lorenzen, J., Surveillance of Vibrations in PWR, *Progress in Nuclear Energy*, Vol. 9, (1982) pp 453–464.

[3] Amazigo, J.C., Buckling of Stochastically Imperfect Structures, in Budiansky, B., ed., *Buckling of Structures,* Springer, Berlin, (1976) pp 172–182.

[4] Anderson, D.L. and Lindberg, H.E., Dynamic Pulse Buckling of Cylindrical Shells Under Transient Lateral Pressure, *AIAA Journal*, Vol. 6(4), (1968) pp 589–598.

[5] Arbocz, J. and Babcock, C.D., Jr., A Multi-Mode Analysis For Calculating Buckling Loads of Imperfect Cylindrical Shells, CALCIT Report SM-74-4, California Institute of Technology, Pasadena, CA, (1974).

[6] Arbocz, J. and Williams, J.G., Imperfection Surveys on a 10 Ft. Diameter Shell Structure, *AIAA Journal*, Vol. 15 (1977), pp 949–956.

[7] Artstein, Z., On the Calculus of Closed Set-Valued Functions, *Indiana University Mathematics Journal*, Vol. 24, pp 433–441 (1974).

[8] Artstein, Z. and Hansen, J.C., Convexification in Limit Laws of Random Sets in Banach Spaces, *Annals of Probability*, Vol. 13, pp 307–309 (1985).

[9] Astil, C.J., Nosseir, S.B. and Shinozuka, M., Impact Loading on Structures With Random Properties, *Journal of Structural Mechanics,* 1: 63 - 77 (1972).

[10] Augusti, G., Stabilita' di Strutture Elastiche Elementari in Presenza di Grandi Spostamenti, (1964), Report No. 172, University of Naples (in Italian).

[11] Aumann, R.J., Integrals of Set-Valued Functions, *Journal of Mathematical Analysis and Applications*, Vol. 12, (1965) pp 1–12.

[12] Balakrishnan, A.V., *Applied Functional Analysis,* 2nd ed., Springer, New York, (1981) p 40.

[13] Basseville, M., Benveniste, A., Moustakides, G., Rougee, A., Optimal Sensor Location For Detecting Changes in Dynamical Systems, 25th IEEE Conference on Decision and Control, Athens, Greece, Dec. (1986) pp 1058 - 1063.

[14] Bedrosian, B., Barbela, M., Drenick, R.F. and Tsirk, A., Critical Excitation Method for Calculating Earthquake Effects on Nuclear Plant Structures: An Assessment Study, NUREG/CR–1673, RD, Burns and Roe, Inc., Oradell, N.J., (1980).

[15] Bekker, M.G., *Introduction to Terrain-Vehicle Systems,* University of Michigan Press, Ann Arbor, (1969).

[16] Bellman, R., *Adaptive Control Processes: A Guided Tour,* Princeton University Press, (1961).

[17] Ben-Haim, Y., *The Assay of Spatially Random Material,* Kluwer Academic Publishers, Dordrecht, The Netherlands, (1985).

[18] Ben-Haim, Y., Convexity Analysis: A Tool for Optimization of Malfunction Isolation, *25th IEEE Conference on Decision and Control*, Athens, Greece, (1986) pp 1570–1575.

[19] Ben-Haim, Y., Detecting Unknown Lateral Forces on a Bar by Vibration Measurement, (1989) to appear in the *Journal of Sound and Vibration.*

[20] Ben-Haim, Y. and Elias, E., Indirect Measurement of Surface Temperature and Heat Flux: Optimal Design Using Convexity Analysis, *International Journal of Heat Mass Transfer,* Vol. 30, (1987) pp 1673–1683.

[21] Ben-Haim, Y. and Elishakoff, I., Non-Probabilistic Models of Uncertainty in the Non-linear Buckling of Shells With General Imperfections: Theoretical Estimates of the Knockdown Factor, *ASME Journal of Applied Mechanics*, (1989a), Vol. 111, pp 403 – 410.

[22] Ben-Haim, Y. and Elishakoff, I., Convex Models of Vehicle Response to Uncertain-But-Bounded Terrain, (1989b), ASME Pressure Vessel and Piping Conference, Honolulu, Hawaii, June, 1989.

[23] Ben-Haim, Y. and Elishakoff, I., Dynamics and Failure of a Thin Bar With Unknown-But-Bounded Imperfections, (1989c), Winter Annual Meeting of the ASME, San Francisco, December, 1989.

[24] Bertsekas, D.P. and Rhodes, I.B., Recursive State Estimation for a Set-Membership Description of Uncertainty, *IEEE Transactions on Automatic Control,* Vol. AC-16, (1971) pp 117–128.

[25] Bieneck, M.P., Fan, T.C. and Lackman, L.M., Dynamic Stability of Cylindrical Shells, *AIAA Journal*, Vol. 4(3), (1966) pp 495–500.

[26] Blekhman, I.I., Myshkis, A.D. and Panovko, Ya. G., *Mechanics and Applied Mathematics: Logics and Specifics of Applications of Mathematics,* "Nauka" Publishing House, Moscow, (1983) (in Russian), pp 205–209.

[27] Bogdanoff, J.L., Cote, L.J. and Kozin, F., Introduction to a Statistical Theory of Land Locomotion - II,*Journal of Terramechanics,* Vol. 2, (1965) No. 3, pp. 17–27.

[28] Bolotin, V.V., Statistical Methods in the Nonlinear Theory of Elastic Shells, *Izvestiya Akademii Nauk SSSR, Otdelenie Tekhnicheskikh Nauk*, No. 3, (in Russian), (1958), English Translation: NASA TTF-85, (1962) pp 1–16.

[29] Bolotin, V.V., Modern Status of the Reliability Theory and Statistical Mechanics of Structures, in *Reliability Problems in Structural Mechanics*, Contributions to the Second USSR Conference on the Reliability Problems in Structural Mechanics, (Bolotin, V.V. and Chyras, A., eds.), "Vaidzas" Publishing House, Vilnius, (1968) (in Russian) pp 7–13.

[30] Boloton, V.V., *Statistical Methods in Structural Mechanics*, Holden Day, San Francisco, (1979).

[31] Braun, S., ed., *Mechanical Signature Analysis: Theory and Applications,* Academic Press, (1986).

[32] Bryson, A.E. and Ho, Y-C, *Applied Optimal Control,* Wiley, New York, (1975).

[33] Budiansky, B. and Hutchinson, J.W., Dynamic Buckling of Imperfection–Sensitive Structures, in Görtler, M., ed., *Proceedings of the 11th International Congress of Applied Mechanics,* Münich, West Germany, (1964) pp 636–651.

[34] Budiansky, B. and Hutchinson, J.W., Buckling: Progress and Challenge, in Besseling, J.F. and Van der Heijden, A.M.A., eds., *Trends in Solid Mechanics,* Sijthoff and Noordhoff International Publishers, Alphen a/d Rijn, The Netherlands, (1979) pp 93–116.

[35] Bulgakov, B.V., Fehleranhaeufung Bei Kreiselapparaten, *Ingenieur-Archiv,* Vol. 11 (1940), pp 461–469 (in German).

[36] Bulgakov, B.V., On the Accumulation of Disturbances in Linear Systems With Constant Coefficients, *Doklady Akademii Nauk SSSR,* Vol. LI, No. 5, (1946), pp 339–342 (in Russian).

[37] Chernous'ko, F.L., Optimal Guaranteed Estimates of Indeterminacies With The Aid of Ellipsoids, *Tekhnicheskaya Kibernetika,* (1981) pp 1–9 (in Russian).

[38] Coppa, A.P. and Nash, W.A., Dynamic Buckling of Shell Structures Subject to Longitudinal Impact, FDL-TDR-64-65, General Electric Corporation, Philadelphia, PA., 19101, December 1964.

[39] Cote, L.J., Kozin, F. and Bogdanoff, J.L., Introduction to a Statistical Theory of Land Locomotion – I, *Journal of Terramechanics*, Vol. 2, (1965) No. 2, pp 17–23.

[40] Cote, L.J., Kozin, F. and Bogdanoff, J.L., Introduction to a Statistical Theory of Land Locomotion – IV: Effects of Vibration on the Contents of the Vehicle and Conclusions, *Journal of Terramechanics*, Vol. 3, (1966) No. 4, pp 47–51.

[41] Davidson, J.F., Buckling of Struts Under Dynamic Loading, *Journal of the Mechanics and Physics of Solids*, Vol. 2, (1953) pp 54–56.

[42] Davies, P. and Hammond, J.K., Envelope and Instantaneous Phase Methods for the Propogation of Transients Through Oscillatory Nonlinear Systems, *ICASSP 86*, (1986) pp 2319–2322, Tokyo.

[43] Delfour, M.C. and Mittler, S.K., Reachability of Perturbed Systems and Min Sup Problems, *SIAM Journal of Control*, Vol. 7, (1969) pp 521–533 (1969).

[44] Drenick, R.F., Functional Analysis of Effects of Earthquakes, *2nd Joint United States –Japan Seminar on Applied Stochastics*, Washington, D.C., (Sept 19 – 24, 1968).

[45] Drenick, R.F., Model-free Design of Aseismic Structures, *Journal of Engineering Mechanics Division*, Proceedings of the ASCE, Vol. 96, No. EM4, (1970), pp 483–493.

[46] Drenick, R.F., On a Class of Non-Robust Problems in Stochastic Dynamics, in B.L. Clarkson, ed., *Stochastic Problems in Dynamics*, Pitman, London, (1977) pp 237–255.

[47] Drenick, R.F., Novomestky, F. and Bagchi, G., Critical Excitation of Structures, in *Wind and Seismic Effects, Proceedings of the 12th Joint UJNR Panel Conference*, US National Bureau of Standards Special Pub., (1984) pp 133–142.

[48] Drenick, R.F. and Yun, C.B., Reliability of Seismic Resistence Predictions, *Journal of the Structural Division*, Proceedings of the ASCE, Vol. 105, (1979) pp 1879–1891.

[49] Duhene-Marullaz, P., Etude des Vitesses Maximales Annuales de Vent, *Cahiers du Centre Scientifique et Technique de Batiment,* No. 131, (1972).

[50] Dym, C.L., *Solid Mechanics: A Variational Approach,* McGraw-Hill, New York (1973).

[51] Efimov, A.B., Malyi, V.I. and Uteshev, S.A., On Loss of Stability of Cylindrical Shell Due to Impact, sl Izvestiya Akademii Nauk SSSR, Mekhanika Tverdogo Tela, No. 1, (1971) pp 20–23 (in Russian).

[52] Elishakoff, I., Axial Impact Buckling of a Column With Random Initial Imperfections, *ASME Journal of Applied Mechanics*, Vol. 45, (1978a) pp 361–365.

[53] Elishakoff, I., Impact Buckling of Thin Bar Via Monte Carlo Method, *ASME Journal of Applied Mechanics,* Vol. 45, (1978b) pp 586–590.

[54] Elishakoff, I., Buckling of a Stochastically Imperfect Finite Column on a Nonlinear Elastic Foundation – A Reliability Study, *ASME Journal of Applied Mechanics,* Vol. 46, (1979) pp 411–416.

[55] Elishakoff, I., Hoff's Problem in a Probabilistic Setting, *ASME Journal of Applied Mechanics,* Vol. 47, (1980) pp 403–408.

[56] Elishakoff, I. and Arbocz, J., Reliability of Axially Compressed Cylindrical Shells With Random Axisymmetric Imperfections, *International Journal of Solids and Structures*, Vol. 18 (1982) pp 563–585.

[57] Elishakoff, I., *Probabilistic Methods in the Theory of Structures,* Wiley-Interscience, New York, (1983a).

[58] Elishakoff, I., How to Introduce the Imperfection–Sensitivity Concept into Design in Thompson, J.M.T. and Hunt, G.W., eds. *Collapse — The Buckling of Structures in Theory and Practice*, Cambridge University Press, Cambridge, England, (1983b) pp 345–357.

[59] Elishakoff, I. and Arbocz, J., Reliability of Axially Compressed Cylindrical Shells With General Nonsymmetric Imperfections, *ASME Journal of Applied Mechanics*, Vol. 52, (1985) pp 122–128.

[60] Elishakoff, I., Van Manen, S., Vermeulen, P.G. and Arbocz, J., First-Order Second-Moment Analysis of the Buckling of Shells With Random Imperfections, *AIAA Journal*, Vol. 25, (1987) pp 1113–1117.

[61] Elishakoff, I. and Ben-Haim, Y., Dynamics of a Thin Cylindrical Shell Under Impact With Limited Deterministic Information on Its Initial Imperfections (1989), European Mechanics Colloquium 250, Nonlinear Mechanical Systems Under Stochastic Conditions, Como, Italy, July 1989.

[62] Etkin, B., The Turbulent Wind and its Effect on Flight, Wright Brothers Lecture, *AIAA Aircraft Systems Meeting*, (August 4–6, 1980) Anaheim, CA., Paper AIAA-80-1836, pp 23.

[63] Fenchel, W., Convexity Through the Ages, in *Convexity and Its Applications*, Gruber, P.M. and Wills, J.M., eds., Birkhäuser, Basel, (1983) pp 120–130.

[64] Feodos'ev, V.I., *Ten Lectures — Conversations on Strength of Materials*, "Nauka" Publishing House, Moscow, (1969) pp 37–40 and 146 (in Russian).

[65] Fraser, W.B. and Budiansky, B., The Buckling of a Column With Random Initial Deflections, *ASME Journal of Applied Mechanics*, Vol. 36, (1969) pp 232–240.

[66] Gerard, G. and Becker, H., Column Behaviour Under Conditions of Impact, *Journal of Aeronautical Sciences*, Vol. 19, (1952) pp 58–65.

[67] Ghiocel, D. and Lungu, D., *Wind, Snow and Temperature Effects on Structures Based on Probability*, Abacus Press, Turnbridge Wells, Kent, U.K. (1975).

[68] Glover, J.D. and Schweppe, F.C., Control of Linear Dynamic Systems With Set Constrained Disturbances, *IEEE Transactions on Automatic Control*, Vol. AC-16, (1971) pp 411–423.

[69] Glover, J.D. and Schweppe, F.C., Advanced Load Frequency Control, *IEEE Transactions*, PAS-91, (1972) pp 2095–2103.

[70] Goldman, S., Vibration Analysis Now Works on Reciprocating Engines, *Power*, (1987) pp 49–51.

[71] Hahn, G. and Shapiro, S., *Statistical Models in Engineering*, Wiley, New York, (1967) pp 73–74.

[72] Hardy, G. H., Littlewood, J.E. and Pólya, G., *Inequalities*, Cambridge University Press, (1934).

[73] Harris, C.M., ed., *Shock and Vibration Handbook*, 3rd edition, McGraw-Hill Book Co., (1987).

[74] Hausner, G.W. and Tso, W.K., Dynamic Behaviour of Supercritically Loaded Struts, *Journal of Engineering Mechanics Division*, Proceedings of the ASCE, (1962) EM5, pp 41–65.

[75] Hessel, G., Liewers, P., Schumann, P., Schmitt, W., and Weiss, F.-P., A Noise Diagnostic System For Operator Advice, *Nuclear Safety*, Vol. 29, (1988) pp 293–306.

[76] Hoff, C. and H.G. Natke, Correction of a Finite-Element Model by Input-Output Measurements with Application to a Radar Tower, *Journal of Modal Analysis*, January, 1989, pp 1 – 7.

[77] Hoff, N.J., The Dynamics of the Buckling of Elastic Columns, *ASME Journal of Applied Mechanics*, Vol. 18, (1951) pp 68–71.

[78] Hoff, N.J., Dynamic Stability of Structures, (Keynote Address), in Herrmann, G. Ed., *Dynamic Stability of Structures*, Pergamon, New York, (1965) pp 7–44.

[79] Huffington, N., Response of Elastic Columns to Axial Pulse Loading, *AIAA Journal*, Vol. 1, (1963) pp 2099–2104.

[80] Iyengar, R.N., Matched Inputs, Report 47, Series J, Center for Applied Stochastics, Purdue University, West Lafayette, IN. (1970).

[81] Jendrzejczyk, J.A., Dynamic Characteristics of Heat Exchanger Tubes Vibrating in a Tube Support Plate Inactive Mode, *ASME Journal of Pressure Vessel Technology,* Vol. 108, (1986) pp 256–266.

[82] Kaiuk, Ya. F., Dynamic Stability of a Bar Due to Axial Impact, *Prikladnaya Mekhanika,* Vol. 1, No. 9, (1965) pp 98–106 (in Russian).

[83] Kelly, P.J. and Weiss, M.L., *Geometry and Convexity: A Study in Mathematical Methods,* Wiley, New York, (1979).

[84] Kiiko, I.A., Cylindrical Shell Under Axial Impact Loading, *Izvestiya Akademii Nauk SSSR, Mekhanika Tverdogo Tela,* No. 2, (1969) pp 135–138 (in Russian).

[85] Kline, M., *Mathematics: The Loss of Certainty,* Oxford University Press, (1980).

[86] Kogan, J., *Crane Design. Theory and Calculations of Reliability,* Wiley, New York (1976), p. 218.

[87] Koiter, W.T., On the Stability of Elastic Equilibrium, Doctoral Thesis, Delft University of Technology, (in Dutch), (1945), English translations, (a) TTF-10, (1967) and (b) AFFDL, TF 70-25, (1970).

[88] Koning, C. and Taub, J., Impact Buckling of Thin Bars in the Elastic Range Hinged at Both Ends, NACA TM 748, (1934) (translation from German).

[89] Kordemski, B.A., *How to Attract Students to Mathematics,* Moscow, (1981) (in Russian).

[90] Kornev, V.M., Modes of Stability Loss in an Elastic Rod Under Impact, *Journal of Applied Mechanics and Technical Physics,* Vol. 9, No. 3, (1968) pp 63–68 (in Russian).

[91] Kornev, V.M., On the Instability Modes of Elastic Shells Under Intensive Loading, *Izvestiya Akademii Nauk SSSR, Mekhanika Tverdogo Tela,* No. 2, (1969) pp 129–135 (in Russian).

[92] Kornev, V.M. and Solodovnikov, V.N., Axially Symmetric Buckling Modes on a Cylindrical Shell Under Impact, *Journal of Applied Mechanics and Technical Physics,* No. 2, (1972) pp 95–100 (in Russian).

[93] Kornev, V.M., Development of the Dynamic Forms of Stability of Elastic Systems Under Intensive Loading on the Finite Time Interval, *Journal of the Applied Mechanics and Technical Physics,* No. 4, (1972) pp 122–128 (in Russian).

[94] Kornev, V.M. and Markin, A.V., Density of Eigenmotions of Elastic Shells Under Intensive Dynamic Loading, *Journal of Applied Mechanics and Technical Physics,* No. 5, (1975) pp 173–178 (in Russian).

[95] Kozin, F. and Bogdanoff, J.L., On the Statistical Analysis of the Motion of Some Simple Two-Dimensional Linear Vehicles Moving on a Random Track, *International Journal of Mechanical Sciences,* Vol. 2, (1960) pp. 168–178.

[96] Kozin, F., Bogdanoff, J.L. and Cote, L.J., Introduction to a Statistical Theory of Land Locomotion – III: Vehicle Dynamics, *Journal of Terramechanics,* Vol. 3, (1966) No. 3, pp. 69–81.

[97] Krick, *An Introduction to Engineering and Engineering Design,* Wiley, New York, (1967).

[98] Kubrusly, C.S. and Malebranche, H., Sensors and Controllers Location in Distributed Systems: A Survey, *Automatica,* Vol. 21, (1985) pp 117–128.

[99] Lavrent'ev, M.A. and Ishlinsky, A.V., Dynamic Forms of Loss of Stability of Elastic Systems, *Doklady Akademii Nauk SSSR,* Vol. 64, (1949) pp 779–782 (in Russian); English translation by R. Cooper, in STL-TR-61-5110-41, Space Technology Laboratories, Los Angeles, CA.

[100] Laws, W.C. and Muszynska, A., Periodic and Continuous Vibration Monitoring For Preventive/Predictive Maintenance of Rotating Machinery, *ASME Journal of Engineering for Gas Turbines and Power,* Vol. 109, (1987) pp 159–167.

[101] Leontief, W., Theoretical Assumptions and Nonobserved Facts, *The American Economic Review,* March, (1971) pp 1–7. Presidential Address to the American Economic Association, Detroit, Michigan, (29 December, 1970).

[102] Lindberg, H.E., Buckling of a Very Thin Cylindrical Shell Due to an Impulsive Pressure, *ASME Journal of Applied Mechanics,* Vol. 31, (1964) pp 267–272.

[103] Lindberg, H.E., Impact Buckling of a Thin Bar, *ASME Journal of Applied Mechanics,* Vol. 32, (1965) pp 312–322.

[104] Lindberg, H.E. and Herbert, R.E., Dynamic Buckling of a Thin Cylindrical Shell Under Axial Impact, *ASME Journal of Applied Mechanics,* Vol. 33, (1966) pp 105–112.

[105] Lindberg, H.E. and Florence, A.L., *Dynamic Pulse Buckling,* Kluwer Academic Publishers, Dordrecht, The Netherlands, (1987) pp 373–380.

[106] Malyshev, V.M., Stability of Columns Under Impact Compression, *Izvestiya Akademii Nauk SSSR, Mekhanika Tverdogo Tela,* Vol. 1, No. 4, (1966) pp 137–142 (in Russian).

[107] Meier, J.H., On the Dynamics of Elastic Buckling, *Journal of Aeronautical Sciences,* Vol. 12, (1945) pp 430–440.

[108] Michalopoulos, C.D. and Riley, T.A., Response of Discrete Linear Systems to Forcing Functions With Inequality Constraints, *AIAA Journal,* Vol. 10, (1972) pp 1016–1019.

[109] NASA, Buckling of Thin-Walled Circular Cylinders, NASA SP-8007, (August, 1968).

[110] Newland, D.E., General Linear Theory of Vehicle Response to Random Road Roughness, in Elishakoff, I. and Lyon, R.H., eds., *Random Vibrations - Status and Recent Developments,* Elsevier Science Publishers, Amsterdam, (1986) pp 303–326.

[111] Nikolaenko, N.A., *Probabilistic Methods of the Dynamic Calculation of the Structures in Mechanical Engineering,* "Mashinostroenie" Publishing House, Moscow, (1967) (in Russian).

[112] Ogata, K., *Modern Control Engineering,* Prentice-Hall, Englewood Cliffs, N.J., (1970) p 774.

[113] Ottestad, P., *Statistical Models and Their Experimental Application,* Griffin, London, (1970) p. 17.

[114] Ozol, J., Check Out Pumps, Valves in Condensate-Piping Vibration, *Power,* Vol. 131, (1987) pp 33–37.

[115] Pickands, J., Asymptotic Properties of Maxima of a Stationary Gaussian Process, *Transactions of the American Mathematical Society,* 145: 75 - 86, (1969).

[116] Pikovsky, A.A., *Statics of Column Systems With Compressed Elements,* Gosudarstvennoe Izdatel'stvo Fiziko-Matemacheskoi Literatury, Moscow, (1961) (in Russian).

[117] Plaut, R.H., Optimal Beam and Plate Foundations For Minimum Compliance, *ASME Journal of Applied Mechanics,* Vol. 54, (1987) pp 255–257.

[118] Popplewell, N. and Youssef, N.A.N., A Comparison of Maximax Response Estimates, *Journal of Sound and Vibration*, Vol. 62, (1979) pp 339–352.

[119] Ricca, P.M. and Bradshaw, P.M., How the Trans-Alaska Pipeline System's Rotating Machinery Maintenance System Evolved, *Oil and Gas Journal,* Vol. 82, No. 33, (1984) pp 96–103.

[120] Robson, J.D., *An Introduction to Random Vibrations,* Edinburgh at the University Press, (1963).

[121] Robson, J.D., Researches in Random Vibrations, in Petyt, M. and Wolfe, H.F., eds., *Recent Advances in Structural Dynamics,* Institute of Sound and Vibration Research, 2nd International Conference, Univ. of Southampton Press, Vol. 2, (1984) pp 621–638.

[122] Robson, J.D., Dodds, C.J., Macvean, D.B. and Paling, V.R., *Random Vibrations,* Courses and Lectures No. 115, Springer, Vienna (1971).

[123] Rockafellar, R.T., *Convex Analysis,* Princeton University Press, Princeton, N.J., (1970).

[124] Rodale, J.I., *The Synonym Finder,* Warner Books, New York, (1978) p 1275.

[125] Roorda, J., *Buckling of Elastic Structures,* Waterloo University Press, Waterloo, Canada, (1980).

[126] Ruiz, P., Discussion of "Maximum Structural Response to Seismic Excitation", *Journal of the Engineering Mechanics Division,* Proceedings of the ASCE, vol. 97, No. EM3, (1971), pp 1024 – 1026.

[127] Schiehlen, W.O., Random Vehicle Vibrations, in Elishakoff, I. and Lyon, R.H., eds., *Random Vibrations – Status and Recent Developments,* Elsevier Science Publishers, Amsterdam, (1986a).

[128] Schiehlen, W.O., Probabilistic Analysis of Vehicle Vibrations, *Probabilistic Engineering Mechanics,* Vol. 1, (1986b) pp 99–104.

[129] Schiehlen, W.O., Analysis and Estimation of Vehicle Systems, in Schiehlen, W.O. and Wedig, W.. eds., *Analysis and Estimation of Stochastic Mechanical Systems,* Springer, Vienna, (1988) pp

[130] Schigley, J.L., *Mechanical Engineering Design,* 3rd ed., McGraw Hill, New York, (1977).

[131] Schlaepfer, F.M. and Schweppe, F.C., Continuous-Time State Estimation Under Disturbances Bounded By Convex Sets, *IEEE Transactions on Automatic Control,* AC-17, (1972) pp 197–205.

[132] Schweppe, F.C., Recursive State Estimation: Unknown But Bounded Errors and System Inputs, *IEEE Transactions on Automatic Control,* Vol. AC-13, (1968) pp 22–28.

[133] Schweppe, F.C., *Uncertain Dynamic Systems,* Prentice-Hall, Englewood Cliffs, N.J., (1973).

[134] Sevin, E., On the Elastic Buckling of Columns Due to Dynamic Axial Forces Including Effects of Axial Inertia, *ASME Journal of Applied Mechanics,* Vol. 27, (1960) pp 125–131.

[135] Shahrivar, F. and Bouwkamp, J.G., Damage Detection in Offshore Platforms Using Vibration Information, *ASME Journal of Energy Resources Technology,* Vol. 108, (1986) pp 97–106.

[136] Shinozuka, M., Maximum Structural Response to Seismic Excitations, *Journal of the Engineering Mechanics Division,* Proceedings of the ASCE, Vol. 96, (1970) No. EM5, pp 729–738.

[137] Simiu, E. and Scanlan, R.H., *Wind Effects on Structures,* Second ed., Wiley, New York, (1986) p 554.

[138] Smith, G.N., *Probability and Statistics in Civil Engineering,* Collins Professional and Technical Books, London, (1986), cover page.

[139] Sobczyk, K. and Macvean, D.B., Non-Stationary Random Vibrations of Systems Travelling With Variable Velocity, in Clarkson, B.L., *et al,* eds., *Stochastic Problems in Dynamics* , Pitman, London, (1976) PP 412–434.

[140] Sobczyk, K., Macvean, D.B. and Robson, J.D., Response to Profile-Imposed Excitation With Randomly Varying Traversal Velocity, *Journal of Sound and Vibration,* Vol. 52, (1977) pp 37–49.

[141] Stoer, J. and Witzgall, C., *Convexity and Optimization in Finite Dimensions,* Springer, Berlin, (1970).

[142] Stumpf, F.B., Demonstration and Laboratory Equipment For Teaching Vibraton and Sound, *Journal of the Acoustical Society of America,* Vol. 82, (1987) pp 1454–1455.

[143] Taub, J., Impact Buckling of Bars in Elastic Range for Any End Conditions, NACA TM 749, 1934 (translation from German).

[144] Thie, J.A., *Power Reactor Noise,* American Nuclear Society, La-Grange Park, USA, (1981).

[145] Thompson, J.M.T. and Hunt, G.W., *A General Theory of Elastic Stability,* Wiley, London, (1973).

[146] Timoshenko, S. P., On the Correction for Shear of the Differential Equation for Transverse Vibrations of Prismatic Bar, *Philosophical Magazine,* Ser 6, 41/245, pp 744 - 746.

[147] Timoshenko, S. P. *Vibration Problems in Engineering,* D. Van Nostrand Company, Toronto (1928).

[148] Timoshenko, S.P. and Gere, J.M., *Theory of Elastic Stability,* McGraw-Hill, Auckland, (1963).

[149] Usher, A.P., *A History of Mechanical Inventions,* Harvard University Press, (1954). Reprinted by Dover Press, (1988).

[150] Vanderplaats, G.N., *Numerical Optimization Techniques for Engineering Design,* McGraw-Hill, New York, (1984), section 5-8.2.

[151] Vanmarcke, E.H., Structural Response to Earthquakes, in C. Lomnitz and E. Rosenblueth, eds., *Seismic Risk and Engineering Decisions,* Elsevier, Amsterdam, (1976), Chapter 8, pp 287 - 337.

[152] Vol'mir, A.S., *Stability of Deformable Bodies,* "Nauka" Publishing House, Moscow, (1967) pp 264–266 (in Russian).

[153] Vol'mir, A.S. and Kil'dibekov, I.G., Investigation of the Buckling Process During Impact, *Doklady Akademii Nauk SSSR,* Vol. 167, No. 4, (1966) (in Russian).

[154] Wang, P.C., Drenick, R.F. and Wang, W., Seismic Assessment of High-Rise Buildings, *Journal of Engineering Mechanics Division,* Proceedings of the ASCE, Vol. 104, No. EM2, (1978), pp 441 - 456.

[155] Webster's Ninth New Collegiate Dictionary, Merriam-Webster Inc., Publishers, Springfield, MA, (1985) p. 1283.

[156] Weinstock, R., *Calculus of Variations,* Dover Publications, New York, (1974) section 4-5.

[157] White, R.G., Acoustic and Vibration Transducers and Measurement Techniques, *Journal of Physics, Part E: Scientific Instruments,* Vol. 18, (1985) pp 790–796.

[158] Wiener, N., *I Am a Mathematician,* MIT Press, Cambridge, MA, (1956).

[159] Williams, M.M.R., *Random Processes in Nuclear Reactors,* Pergamon, (1974).

[160] Witsenhausen, H.S., A Minimax Control Problem For Sampled Linear Systems, *IEEE Transactions on Automatic Control,* Vol. AC-13, (1968a) pp 5–21.

[161] Witsenhausen, H.S., Sets of Possible States of Linear Systems Given Perturbed Observations, *IEEE Transactions on Automatic Control,* Vol. AC-13, (1968b) pp 556–558.

[162] Young, D.M., Stresses in Eccentrically Loaded Steel Columns, Publication of the International Association of Bridge Structural Engineers, Vol. 1 (1932).

[163] Youssef, N.A.N. and Popplewell, N., The Maximax Response of Discrete Multi-Degree-of-Freedom Systems, *Journal of Sound and Vibration*, Vol. 64, (1979) pp. 1–15.

[164] Zincik, D.G. and Tennyson, R.C., Stability of Circular Cylindrical Shells Under Transient Axial Impulsive Loading, *AIAA Journal,* Vol. 18, (1980) pp 691–699.

Index